2023北京国际首饰艺术展

2023 Beijing International Jewellery Art Exhibition

主　编｜詹炳宏　　　　副主编｜高伟｜胡俊

EDITOR｜ZHAN BINGHONG　　DEPUTY EDITORS｜GAO WEI｜HU JUN

中国纺织出版社有限公司

内 容 提 要

自2013年起，北京国际首饰艺术展已经成功举办五届，是迄今为止全球规模最大的首饰艺术学术展览之一。2023北京国际首饰艺术展以"美美与共"为主题，强调不同文化之间的取长补短、相互促进，展现首饰艺术的精妙之境，打破首饰艺术的界限壁垒，展览旨在搭建一个多元的首饰文化交流平台，推进多元化首饰艺术创作的潮流，解锁首饰艺术创作的更加广阔的空间。本届展览分为五个部分："数字·域""工艺·域""材料·域""观念·域"和"未来·域"，分别代表了当下珠宝首饰设计学术领域发展的不同方向。展览共收到来自全球25个国家135位艺术家的作品，企业展区共展出国内珠宝首饰领域最著名的20多家企业的展品，以及"未来·域"设计大赛的获奖作品，全部展品共计800余件。

本书图文并茂，图例丰富，适合高等院校珠宝首饰专业师生、珠宝首饰设计师、收藏家以及广大首饰爱好者阅读与参考。

图书在版编目（CIP）数据

2023北京国际首饰艺术展 / 詹炳宏主编；高伟，胡俊副主编 . -- 北京：中国纺织出版社有限公司，2025.5. -- ISBN 978-7-5229-2362-8

Ⅰ . TS934.3

中国国家版本馆 CIP 数据核字第 2024U1H576 号

责任编辑：李春奕 　　特约编辑：徐铭爽
责任校对：高 涵 　　责任印制：王艳丽

中国纺织出版社有限公司出版发行
地址：北京市朝阳区百子湾东里 A407 号楼 　邮政编码：100124
销售电话：010—67004422 　传真：010—87155801
http://www.c-textilep.com
中国纺织出版社天猫旗舰店
官方微博 http://weibo.com/2119887771
北京华联印刷有限公司印刷 　各地新华书店经销
2025 年 5 月第 1 版第 1 次印刷
开本：787×1092 　1/8 　印张：31
字数：150 千字 　定价：468.00 元

FLOURISH
AND
PROSPER
TOGETHER

2023北京国际首饰艺术展
2023 Beijing International
Jewellery Art Exhibition

2023 北京国际首饰艺术作品展

展览汇聚了来自全球 25 个国家的 135 位艺术家与设计师制作的共计 800 余件作品。展览分为"数字·域""工艺·域""材料·域""观念·域"和"未来·域"五个部分，分别代表了当下珠宝首饰设计学术领域发展的不同方向。展览时间为 2023 年 10 月 14 日 ~ 10 月 22 日。

北京国际首饰艺术展开幕式及参展作品秀

区别于国际首饰艺术静态展，本次展览以动态时尚走秀的形式呈现世界各国艺术家及设计师的首饰作品。

"未来·域"珠宝首饰设计大赛

"未来·域"珠宝首饰设计大赛旨在弘扬珠宝首饰理论，推动创新发展，聚合来自创新、技术、智造文化等多层次、多维度的设计指向，让传统与时尚、传承与创新在这里同频共振，深挖珠宝行业的潜力，积极探索时尚与未来生活的平衡，使优秀的设计获得更多的社会关注，通过设计师的演绎，开创我国珠宝首饰行业新时代的辉煌。肩负使命，向新而行。

全球首饰设计教育峰会

针对首饰艺术教育话题，分析国际语境下的首饰艺术，探讨商业首饰的发展方向、趋势和未来首饰设计人才的需求。

第六届北京国际首饰艺术展高峰论坛

该论坛汇聚了具有影响力的首饰专家学者、产业精英和珠宝企业家，是在中国境内举办的最高水平的国际首饰艺术与设计学术活动。

2023 Beijing International Jewellery Art Exhibition

The exhibition brings together over 800 works by 135 artists and designers from 25 countries and regions around the world. The exhibition is divided into five parts: "Digital · Field" "Craftsmanship · Field" "Materials · Field" "Concepts · Field" and "Future · Field", representing different directions in the current academic field of jewellery design. The exhibition is from October 14th to October 22nd, 2023.

Opening Ceremony and Exhibition Works Show of Beijing International Jewellery Art Exhibition

Unlike international static exhibitions of jewellery art, this exhibition presents jewellery works by artists and designers from around the world in the form of dynamic fashion shows.

Future Field Jewellery Design Competition

The purpose of the Future Field Jewellery Design Competition is to promote jewellery theory, promote innovative development, aggregate design directions from multiple levels and dimensions, such as innovation, technology, and intelligent manufacturing culture, resonate with tradition and fashion, inheritance and innovation, deeply cultivate the potential of the jewellery industry, actively explore the balance between fashion and future life, and attract more social attention to excellent designs. Through the interpretation of designers, create a new era of brilliance in China's jewellery industry. Let us take the mission, march toward the new era.

Global Jewellery Design Education Summit

Regarding the topic of jewellery art education, this summit explores the development direction and trends of jewellery art and commercial jewellery in the international context, as well as the development needs of jewellery design talents in the future.

The 6th Beijing International Jewellery Art Exhibition Summit Forum

This forum brings together influential jewellery experts, scholars, industry elites, and jewellery entrepreneurs, and is the highest level international jewellery art and design academic event held in China.

本次展览的主题为"美美与共",面对当下珠宝首饰设计学术领域发展的不同方向,推进多元化,探讨新观念,以观念创新、学术引导为主,为不同的艺术创作观念和设计思潮提供互动的平台与空间。

The theme of this exhibition is " Flourish and Prosper Together ". Faced with the different directions of development in the academic field of jewellery design, promoting diversification, exploring new concepts, and focusing on conceptual innovation and academic guidance, we aim to build an interactive platform and space for different artistic creation concepts and design trends.

组织方式

★✧✧✦✦✧

指导单位：

中国工艺美术学会

工业和信息化部工业文化发展中心

Ico-D 国际设计理事会

主办单位：

北京服装学院

承办单位：

服饰艺术与工程学院

中国生活方式设计研究院

国际首饰设计高校联盟

协办单位：

北京工艺美术学会

北京设计学会

《艺术设计研究》杂志社

《设计》杂志社

学术支持：

《艺术设计研究》

Klimt02

《设计》

《中国宝石》

《中国珠宝首饰》

《中国黄金珠宝》

《芭莎珠宝》

《中国黄金报》

支持单位：

外国院校：

英国皇家艺术学院

英国伦敦艺术大学

英国布莱顿大学

英国曼彻斯特都市大学

英国伯明翰城市大学

意大利佛罗伦萨阿契米亚首饰学院

意大利佛罗伦萨欧纳菲珠宝设计学院

意大利帕多瓦塞瓦蒂克国立艺术学院

比利时安特卫普皇家美术学院

德国慕尼黑美术学院

德国普福尔茨海姆应用技术大学

西班牙巴塞罗那玛萨纳艺术学院

美国罗德岛设计学院

日本东京艺术大学

南非德班理工大学

韩国首尔大学

韩国国民大学

韩国弘益大学

中国院校：

清华大学美术学院

中央美术学院

中国地质大学

中国美术学院

南京艺术学院

上海大学美术学院

天津美术学院

四川美术学院

广州美术学院

山东工艺美术学院

鲁迅美术学院

北京工业大学艺术设计学院

企业公司：

北京菜市口百货股份有限公司

北京珐琅厂有限责任公司

北京玉尊源玉雕艺术有限责任公司

萃华金银珠宝股份有限公司

国金有限公司

上海老凤祥有限公司

上海豫园黄金珠宝集团有限公司

上海铂利德钻石有限公司

上海华泰珠宝商场有限公司

上海张铁军珠宝集团有限公司

南京通灵珠宝股份有限公司

周大福珠宝金行有限公司

周生生珠宝金行有限公司

周大生珠宝股份有限公司

广东潮宏基实业股份有限公司

信德缘集团有限公司

深圳市百泰珠宝首饰有限公司

深圳市粤豪珠宝有限公司

深圳市甘露珠宝首饰有限公司（爱得康）

深圳盛峰黄金有限公司

深圳市沃尔弗斯实业有限公司

中和盛世珠宝文化发展有限公司

深圳峰汇珠宝首饰有限公司

国泉金业（北京）文化股份有限公司

北京和玉缘和田玉珠宝有限公司

ORGANIZATION

Steering organization :

China National Arts & Crafts Society (CNACS)
Industrial Culture Development Center of Ministry
of Industry and Information Technology (MIIT)
International Council of Design (Ico-D)

Sponsor :

Beijing Institute of Fashion Technology (BIFT)

Organizer:

School of Fashion Accessory
Chinese Academy of Lifestyle Design
International Jewellery College Association

Co-organizers:

Beijing Arts and Crafts Association
Beijing Design Society
Art & Design Research
Design

Academic support:

Art and Design Research
Klimt02
Design
China Gems

China Jewelry
Jewelery & Gold
Bazaar Jewelry
China Gold News

Supporting organizations :

Foreign Colleges and Universities:
Royal College of Art (UK)
University of the Arts London (UK)
University of Brighton (UK)
Manchester Metropolitan University (UK)
Birmingham City University (UK)
Alchimia Contemporary Jewellery School
(Florence, Italy)
Le Arti Orafe Jewellery School & Academy
(Florence, Italy)
Pietro Selvatico Art Institute (Padua, Italy)
Royal Academy of Fine Arts
(Antwerp, Belgium)
Academy of Fine Arts (Munich, Germany)
Pforzheim University of Applied Sciences
(Germany)
Massana School of Art and Design
(Barcelona, Spain)
Rhode Island School of Design (USA)
Tokyo University of the Arts (Japan)
Durban University of Technology (South Africa)
Seoul National University (South Korea)
Kookmin University (South Korea)
Hongik University (South Korea)

China's Colleges and Universities:
Academy of Arts & Design
Tsinghua University
Central Academy of Fine Arts
China University of Geosciences
China Academy of Art
Nanjing University of the Arts
Shanghai Academy of Fine Arts
Tianjin Academy of Fine Arts
Sichuan Fine Arts Institute
Guangzhou Academy of Fine Arts
Shandong University of Art & Design
Luxun Academy of Fine Arts
College of Art and Design
Beijing University of Technology

Enterprise Companies:
Beijing Caishikou Department Store Co., Ltd.
Beijing Enamel Factory Co., Ltd.
Beijing Yu Zun Yuan Jade Sculpture Co., Ltd.
Cuihua Gold, Silver and Jewellery Co., Ltd.
Guojin Gold Co., Ltd.
Shanghai Lao Feng Xiang Co., Ltd.
Shanghai Yuyuan Gold and Jewellery Group Co., Ltd.

Shanghai Bolide Diamond Co., Ltd.
Shanghai Huatai Jewellery Department Store Co., Ltd.
Shanghai Zhang Tie Jun Jewellery Group Co., Ltd.
Nanjing Tesiro Jewellery Co., Ltd.
Chow Tai Fook Jewellery Group Co., Ltd.
Chow Sang Sang Jewellery Co., Ltd.
Chow Tai Seng Jewellery Co., Ltd.
Guangdong CHJ Jewellery Co., Ltd.
Xin De Yuan Group Co., Ltd.
Shenzhen Batar Jewellery Co., Ltd.
Shenzhen Yuehao Jewelery Co., Ltd.
Shenzhen Ganlu Jewelry Co., Ltd. (ADK)
Shenzhen Sunking Jewelry Co., Ltd.
Shenzhen Wolfers Industrial Co., Ltd.
Zhong He Sheng Shi Jewellery Culture Development Co., Ltd.
Shenzhen Foreway Jewellery Group Co., Ltd.
Guoquan Gold (Beijing) Culture Co., Ltd.
Beijing HYY Hotan Jade & Jewellery Co., Ltd.

组织委员会

主席：

贾荣林　北京服装学院　校长

秘书长：

詹炳宏　北京服装学院　副校长

组委会办公室：

主任：崔敏

副主任：高伟　闫磊　陈行斌　申利华

成员：张弘　胡俊　邹宁馨　傅永和

赵祎　潘峰　宋懿　熊芊芊　唐天

程之璐　刘小奇　王涛　韩欣然

王浩睿　付少雄　国情　李昕

杨思容　吴青蔓　周晓童　杨子涵

崔艺铭　单怡宁　魏勤文　刘源

贺爽　张弛　张帆　苏艺　宋佳珈

委员：

白静宜　包晓莹　陈火龙　储卫民

才大颖　程学林　陈国珍　陈晓华

常炜　杜半　范海民　韩雨蒙

郭强　郭颖　郭英杰　郭新　高伟

洪兴宇　胡书刚　黄雯

Laurent-Max De Cock（比利时）

Leo Caballero（西班牙）　兰翠芹

李春珂　李峻　李秀美　李英杰

李雪梅　刘骁　廖创宾　马世忠

Maria Rosa Franzin（意大利）　任进

Peter Deckers（新西兰）　潘团结

齐红　宋处岭　施垒　施健

孙仲鸣　唐绪祥　滕菲　王春利

王春刚　王晓昕　王志伟　汪正虹

文乾刚　许平　许梦佳　袁桂平

叶向洲　月文　张纯辉　张福文

张铁成　张世忠　张卫峰　赵丹绮

郑静　郑裕彤　郑耿坚　钟连盛

周桃林　周厚厚　周宗文　朱伟明

庄冬冬　邹宁馨　毕立君　孙凤民

刘江毅　林旭东　谢昭华

执行委员：

韩欣然　胡俊　傅永和　赵祎

潘峰　宋懿　熊芊芊　唐天　程之璐

刘小奇　王涛　王浩睿　付少雄

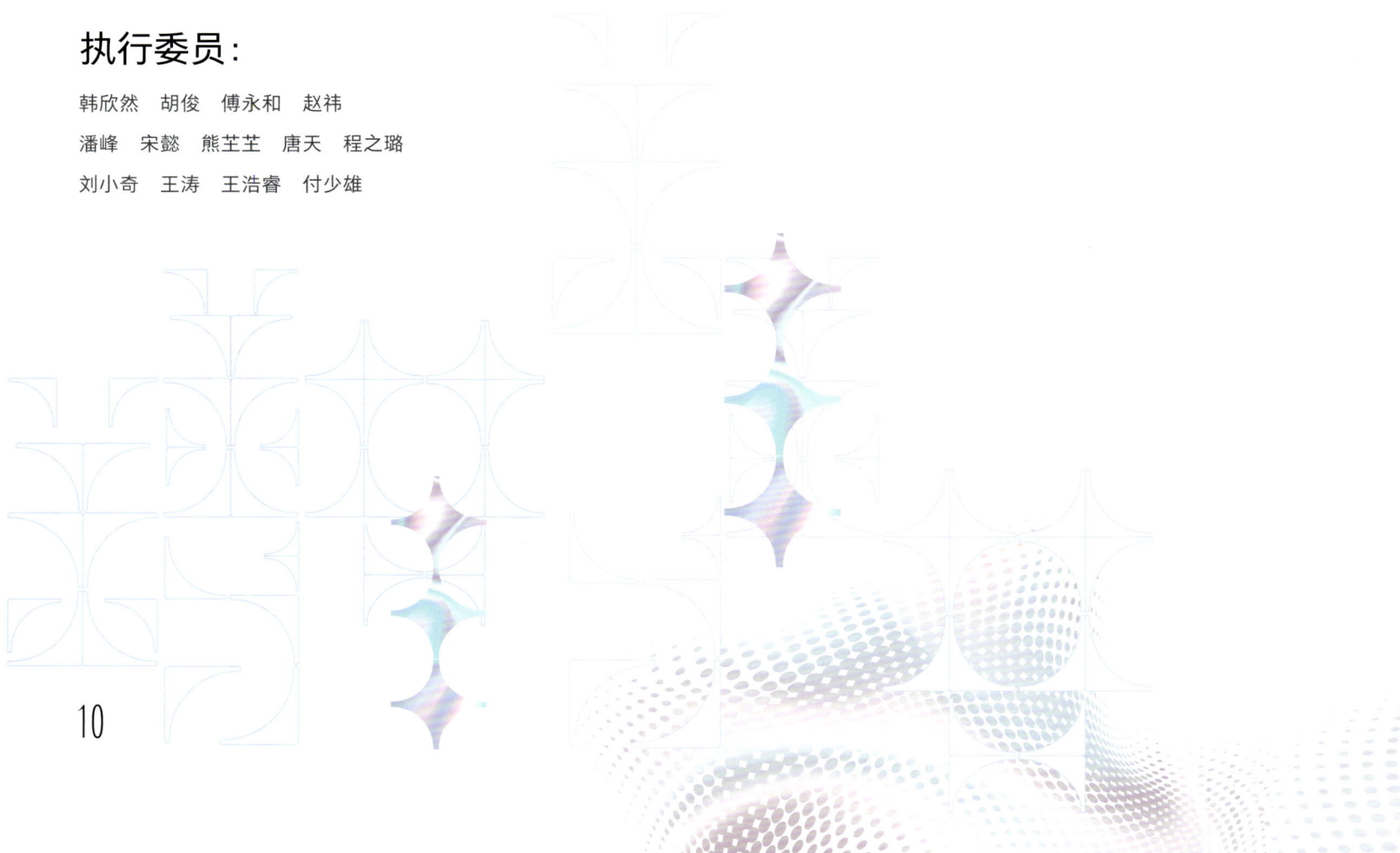

ORGANIZING COMMITTEE

Chairperson:

Jia Ronglin, President of BIFT

Secretary-General:

Zhan Binghong, Vice President of BIFT

Organizing Committee Office:

Directors: Cui Min
Vice Directors: Gao Wei, Yan Lei, Chen Xing-bin, Shen Lihua
Members: Zhang Hong, Hu Jun, Zou Ningxin, Fu Yonghe, Zhao Yi, Pan Feng, Song Yi, Xiong Dudu, Tang Tian, Cheng Zhilu, Liu Xiaoqi, Wang Tao, Han Xinran, Wang Haorui, Fu Shaoxiong, Guo Qing, Li Xin, Yang Sirong, Wu Qingman, Zhou Xiaotong, Yang Zihan, Cui Yiming, Shan Yining, Wei Qinwen, Liu Yuan, He Shuang, Zhang Chi, Zhang Fan, Su Yi, Song Jiajia

Committee Member:

Bai Jingyi, Bao Xiaoying, Chen Huolong, Chu Weimin, Cai Daying, Cheng Xuelin, Chen Guozhen, Chen Xiaohua, Chang Wei, Du Ban, Fan Haimin, Han Yumeng, Guo Qiang, Guo Ying, Guo Yingjie, Guo Xin, Gao Wei, Hong Xingyu, Hu Shugang, Huang Wen, Laurent-Max De Cock (Belgium), Leo Caballero (Spain), Lan Cuiqin, Li Chunke, Li Jun, Li Xiumei, Li Yingjie, Li Xuemei, Liu Xiao, Liao Chuangbin, Ma Shizhong, Maria Rosa Franzin (Italy), Ren Jin, Peter Deckers (New Zealand), Pan Tuanjie, Qi Hong, Song Chuling,

Shi Kun, Shi Jian, Sun Zhongming, Tang Xuxiang, Teng Fei, Wang Chunli, Wang Chungang, Wang Xiaoxin, Wang Zhiwei, Wang Zhenghong, Wen Qiangang, Xu Ping, Xu Mengjia, Yuan Guiping, Ye Xiangzhou, Yue Wen, Zhang Chunhui, Zhang Fuwen, Zhang Tiecheng, Zhang Shizhong, Zhang Weifeng, Zhao Danqi, Zheng Jing, Zheng Yutong, Zheng Gengjian, Zhong Liansheng, Zhou Taolin, Zhou Houhou, Zhou Zongwen, Zhu Weiming, Zhuang Dongdong, Zou Ningxin, Bi Lijun, Sun Fengmin, Liu Jiangyi, Lin Xudong, Xie Chaohua

Executive Committee:

Han Xinran, Hu Jun, Fu Yonghe, Zhao Yi, Pan Feng, Song Yi, Xiong Dudu, Tang Tian, Cheng Zhilu, Liu Xiaoqi, Wang Tao Wang Haorui, Fu Shaoxiong

2023

北京国际首饰艺术展
· 美美与共

2023

Beijing International Jewellery
Art Exhibition
· Flourish and
Prosper Together

✦✦✦

序

★◆☆☆◆◇

首饰的历史大概可以追溯到遥远的石器时代。首饰因宗教功能动因和社会功能动因而起源和发展。伴随着人类历史的发展进程，首饰的原材料、佩戴人群、象征意义和作用都在日益丰富。首饰在时代变迁中历久弥新，向今世人诉往昔事。中外首饰领域的设计师、匠人、从业者们为满足人民日益提升的精神审美和物质需求，不断地求索探讨创新。伴随着生活水平的提高，首饰在人与人之间、文明之间的沟通、交流中发挥着独特的作用。开展首饰领域的研究交流和创新，有利于珠宝产业的创新发展和相关领域人才的培养，有助于增进不同民族和文化之间的交流理解和欣赏。北京服装学院坚持"求实创新、学以致用"的办学理念，形成了"以艺为主、服装引领、艺工融合"的办学特色，建设了中关村服饰时尚设计产业创新园。民族服饰博物馆是中国第一家服饰类专业博物馆。北京服装学院以独特的办学优势和鲜明的办学特色，富有成效的服装服饰教育实践，对我国服装设计时尚和文化创意人才培养，以及产业发展做出了突出贡献。在中华优秀传统文化的传承和弘扬方面发挥了重要作用。北京服装学院依托办学优势与教学特色，搭建北京国际首饰艺术展这个国际学术平台，既体现了提高人才培养质量的更高追求，也提供了促进人文交流的实践平台。

2023
北京国际首饰艺术展
组委会

PREFACE

The history of jewellery can probably be traced back to the remote stone age. The originated and developed jewellery due to religious and social function drivers. With the progress of human history, the raw materials, wearers, symbolic significance and functions of jewellery are increasingly enriched. With the changes of the times, jewellery has been renewed over the years, showing the world about its history. Designers, artisans and practitioners in the field of jewellery at home and abroad constantly seek, explore and innovate to meet the people's increasing spiritual aesthetic and material needs. With improved living standards, jewellery plays a unique role in the communication and exchange between people and civilizations. Carrying out research and innovation exchange in the jewellery field is conducive to the innovative development of the jewellery industry and the cultivation of talents in related fields, and helps to enhance the communication, understanding and appreciation among different nationalities and cultures. Beijing Institute of Fashion Technology adheres to the educational philosophy of seeking innovation and applying the learning to practical problems, forming the school running characteristics of clothing leading the integration of art and industry, and building the BIFT park. The Museum of national costumes is the first professional museum of costumes in China. With its unique advantages, distinctive school running characteristics and fruitful clothing education practice, Beijing Institute of Fashion Technology has made unique and outstanding contributions to the training of fashion and cultural creative talents in China's clothing design and industrial development. It has played an important role in inheriting and carrying forward the excellent traditional Chinese culture. Beijing Institute of Fashion Technology relies on its advantages and teaching characteristics. The holding of Beijing International Jewellery Art Exhibition, an international academic platform, improve the quality of talent training, and built a practical platform to promote people-to-people and cultural exchanges.

2023
Beijing International Jewellery Art Exhibition
Organizing Committee Office

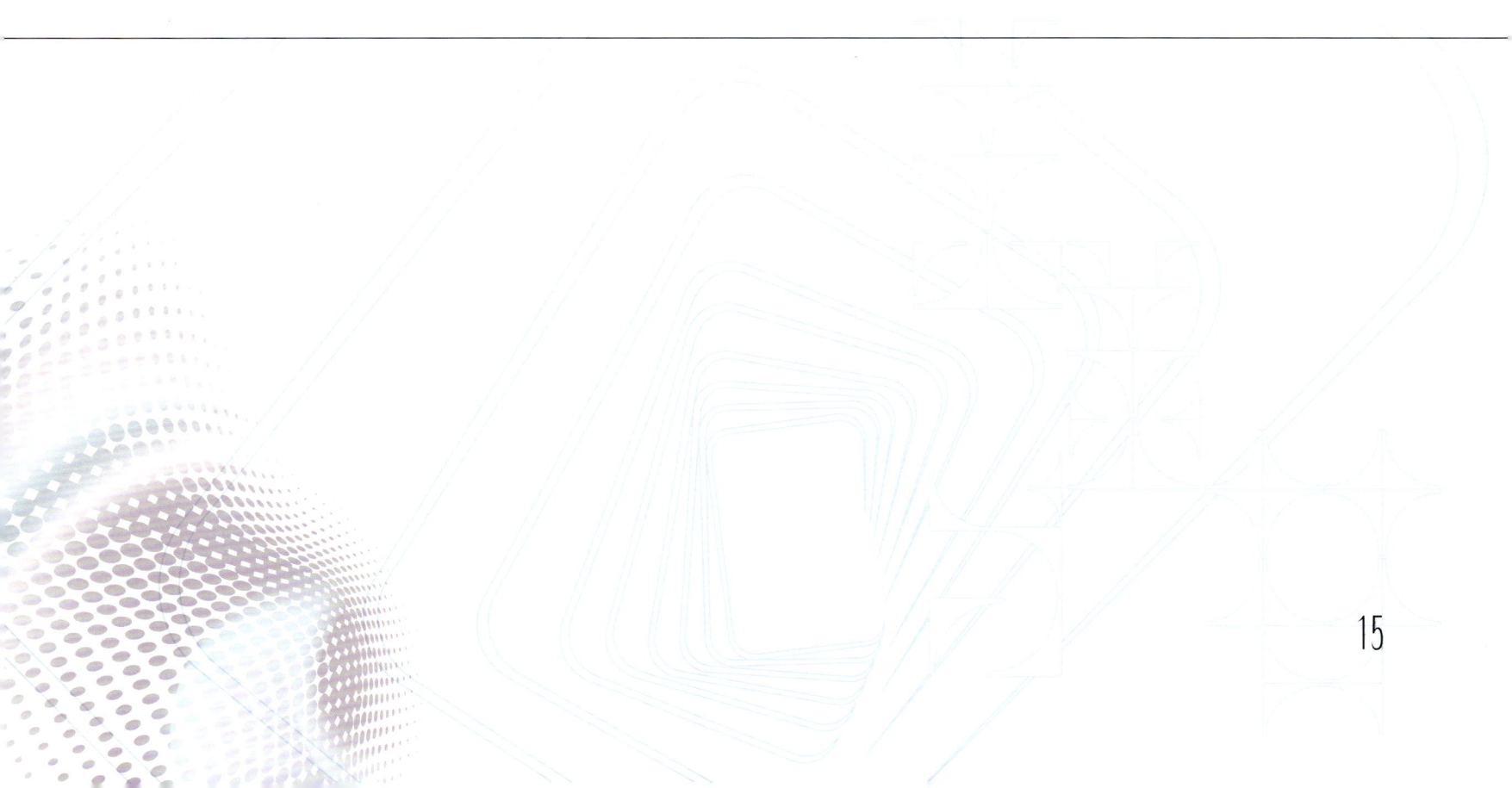

致辞 1

◆◇◇◇◆◇

让我们共同领略一下今天展现在我们面前的精彩时刻。这个展览是对首饰艺术的致敬，也是对设计和工艺的高度赞誉。在这里，您将会看到来自世界各地杰出设计师的作品，他们通过巧夺天工的设计和精湛的工艺，将首饰艺术推向了新的高度。

本次展览的主题是"美美与共"，我们希望通过这个主题，突显首饰作为一门艺术形式的多样性和包容性。在这些珠宝背后，无数设计者在探索不同文化、传统和工艺，让我们的心灵在这艺术的海洋中畅游。首饰不仅是美丽的点缀，更是一种表达个性、传递情感的媒介。每一件作品都是设计师对世界的独特解读，是时间、历史和文化的交汇点。

在这里，我们能够感受到全球不同地域、不同文化背景的设计灵感的碰撞和交融。这不仅是一场视觉的盛宴，更是一次文化的交流与碰撞，通过这些艺术品，我们能够共同见证世界各地首饰艺术的传统、历史，及其与首饰设计者的表达、创新之融合。

最后，感谢所有支持这次活动的设计师、工匠、策展人、工作人员以及每一位到场的嘉宾和观众朋友。让我们共同沉浸在美的海洋中，愿这次展览成为一个启迪心灵、感受多元文化的独特平台。

谢谢！祝愿大家在本次展览中度过一段难忘的时光！

周志军

北京服装学院党委书记

SPECH 1

Let's enjoy the wonderful moment presented before us today. This exhibition is a tribute to jewellery art and a high praise for the design and craftsmanship. Here, you will see masterpieces from outstanding designers around the world who have taken jewellery art to new heights through their exquisite designs and craftsmanship.

The theme of this exhibition is " Flourish and Prosper Together ", and we hope to highlight the diversity and inclusiveness of jewelry as an art form through this theme. Behind these jewellery, countless designers are exploring the exquisite craftsmanship of different cultures, traditions, and materials, allowing our hearts to swim freely in this ocean of art. Jewellery is not only a beautiful decoration but also a medium for expressing personality and conveying emotions. Every piece of work is a unique interpretation of the world by the designer, and it is the intersection of time, history, and culture.

Here, we can feel the collision and fusion of design inspirations from different regions and cultural backgrounds around the world. This is not only a visual feast but also a cultural exchange and collision. Through these artworks, we have witnessed the traditions and history of jewellery art from all over the world, as well as the integration of expression and innovation with jewellery designers.

Finally, thank you to all the designers, craftsmen, curators, staff who have supported this event, as well as every guest and audience present. May this exhibition become a unique platform for us to immerse ourselves in the ocean of beauty, inspire our souls and experience diverse cultures together.

Thank you! Wishing everyone an unforgettable time in this exhibition!

Zhou Zhijun

Party Committee Secretary of
Beijing Institute of Fashion Technology

致辞 2

★☆☆◆◆◇

今天是 2023 北京国际首饰艺术展高峰论坛举办的日子。自 2013 年起，北京国际首饰艺术展及高峰论坛已经成功举办五届，是迄今为止全球规模最大的首饰艺术学术展览。

北京服装学院以建设国际一流时尚高校为目标，多年来面向时尚和文化创意产业，逐渐成为人才培养与时尚创新的高地。北京国际首饰艺术展和国际首饰艺术高峰论坛创建平台，进行高峰对话、学术交流，在国际语境中共同探讨首饰艺术的未来，是高校实践学术报国、服务人民追求美好生活需要的内在要求。"高校立身之本在于立德树人。只有培养出一流人才的高校，才能够成为世界一流大学。"一所大学办得好不好，不是看它的条件何等优越、规模如何庞大，而是要以长远、历史的视野看它培养出什么样的人才，看它对国家对民族所做的贡献。北服一直以来源源不断地为中国纺织服装、时尚和文化创意产业培养输送人才，一直把立德树人作为学校的根本任务，立志培养出对国家、对社会、对人民的有用之才。作为中国时尚高等教育的领军人，我们始终坚持立足首都、面向全国，在中国时尚教育教学科研中发挥引领和先驱作用。北京国际首饰艺术展和高峰论坛作为一项重要内容，引入学术资源和时尚力量，实现国际对话。

本次展览的主题为"美美与共"，面对当下珠宝首饰设计学术发展的不同方向，推进多元化，探讨新观念，以观念创新、学术引导为主，为不同的艺术创作观念和设计思潮搭建互动的平台与空间。

首饰作为一种身体装饰，代表着人类对美的追求，更承载了人的思想与情感，铭刻了源远的文化基因。行至多元化发展的今天，首饰设计不仅继续在原有领域孜孜不倦地探索，更将目光转向新的形势与媒介、技艺与手段，探索科技、精研技艺、延伸交互与关系，显示出更全方位的介入。本届论坛旨在打造首饰艺术设计全球对话平台，对于国际首饰行业的发展动态与时尚流行趋势、新技术及人工智能浪潮席卷下首饰领域的未来、可持续设计观念如何影响未来首饰教育的发展、全球化背景下如何开展首饰人才培养等重要问题皆有思考和讨论。

今天嘉宾云集、高朋满座，不仅带来全新的观念和权威的学术引导，更是一场观点的碰撞与思想的盛宴。预祝 2023 北京国际首饰艺术高峰论坛圆满成功！

贾荣林

北京服装学院校长

SPEECH 2

Today is the day of the 2023 Beijing International Jewellery Art Exhibition Summit Forum. Since 2013, the Beijing International Jewellery Art Exhibition and Summit Forum have been successfully held for five sessions, making it the most prominent jewellery art academic exhibition in the world to date.

Beijing Institute of Fashion Technology aims to build an internationally renowned university on fashion and has been focusing on the fashion and cultural creative industries for many years, gradually becoming a highland for talent cultivation and fashion innovation. The Beijing International Jewellery Art Exhibition and the International Jewellery Art Summit Forum, creating platforms for peak dialogues and academic exchanges, and jo ntly exploring the future of jewellery art in the international context are an intrinsic requirement for universities to provide academics to serve the country and to serve the needs of the people in pursuit of a better life. Beijing Institute of Fashion Technology has always been continuously cultivating and delivering talents for China's textile and clothing, fashion, and cultural industries. It has always regarded cultivating virtue and talent as the fundamental task of the school, and is committed to cultivating valuable talents for the country, society, and people. As a leading figure in China's fashion higher education, we always adhere to the principle of being based in the capital and facing the whole country, and play a leading and pioneering role in the teaching and research of fashion education in China. As important content, the Beijing International jewellery Art Exhibition and Summit Forum, introduce academic resources and fashion power into international dialogue.

The theme of this exhibition is " Flourish and Prosper Together ". Faced with the different directions of development in the academic field of jewellery design, promoting diversification, exploring new concepts, and focusing on conceptual innovation and academic guidance, we aim to build an interactive platform and space for different artistic creation concepts and design trends.

Jewellery, as a long-standing form of body decoration, represents the human pursuit of beauty, carries human thoughts and emotions, and engraves cultural genes from a distant source. In today's diversified development world, jewellery design not only continues to tirelessly explore its original field, but also turns its attention to new situations and media, skills and means, exploring technology, finely researching skills, extending interaction and relationships, demonstrating more comprehensive intervention. This forum aims to create a global dialogue platform for jewellery art and design, to discuss the development trends and fashion trends of the international jewellery industry; the future of the jewellery industry is swept by the wave of new technologies and artificial intelligence; how sustainable design concepts affect the development of future jewellery education; there are important issues such as how to carry out jewellery talent training in the context of globalization that have been pondered and discussed.

Today's gathering of guests and distinguished guests not only brings new ideas and authoritative academic guidance but also a feast of collision of viewpoints and ideas. Wishing the 2023 Beijing International Jewellery Art Summit a complete success!

Jia Ronglin

President of
Beijing Institute of Fashion Technology

致辞 3

首饰在中国工艺美术众多品类中占有独特的历史地位，既是一个独立的艺术门类，又与其他门类的发展有着千丝万缕的联系，这一特质赋予了首饰无穷的时代魅力。首饰作为身体的装饰，铭刻着源远流长的文化基因，承载着人类对于美的追求与情感。首饰始于人类对自然的崇拜，与人类文明起源同步，人类通过身体装饰与自然的对话，赋予首饰以精神内涵。其所蕴含的民族、人文、地域、历史、自然、思想、情感等种种要素，凝聚成了源远流长的首饰设计文化，因而在工艺美术门类中独树一帜。

北京服装学院以建设国际一流时尚高校为目标，多年来面向时尚和文化创意产业，逐步发展成为这一领域人才培养与时尚创新的高地。北京国际首饰双年展和国际首饰艺术高峰论坛已经在业内小有影响力，起到建构平台、高峰对话、学术交流等作用，北京服装学院为业内同仁在国际语境中共同探讨首饰艺术的未来铺就了一条绿色通道。既是责任使然，更是使命担当，表现出了一所追求一流时尚高校的格局与境界，落地有声地践行着学术报国、服务人民追求美好生活需要的办学宗旨。

本次展览的主题为"美美与共"，强调不同文化之间取长补短、相互促进，通晓首饰艺术的精妙之境，打破首饰艺术的壁垒，实乃工艺美术行业之幸事，在首饰发展史上当可大书一笔。2023 北京国际首饰艺术展彰显出"多元化"的力量，旨在搭建一个多元首饰文化交流的平台，以此推进多元化首饰艺术创作的潮流，为首饰艺术创作提供更加广阔的提升空间。面向需求个性化的未来，今日之首饰，已经不仅仅是提高生活品质的重要艺术媒介，文化载体的属性与功能日渐突出。首饰不断突破固有的形态，设计者不止于诠释首饰的独特魅力，更不忌惮于探索新的形式与媒介、技艺与手段，首饰设计呈现出多元化的创新态势是时代之需，我们要做的就是顺势而为，满足时代需求。

当然，展览期间会有很多议题需要同仁们开诚布公地发表己见，包括但不限于这样一些主题："在艺术与时尚不断交融的今天，艺术的时尚化与商业化无须回避现象""国际珠宝商业的发展动态与时尚流行趋势作为首饰艺术发展要面临的问题""面对人工智能浪潮的到来，首饰领域有可能发生的根本性改变"以及"可持续设计观念如何影响未来首饰教育的发展""全球化背景下如何开展首饰人才培养"等业内外都十分关注的问题。我相信，在与会作者与专家同仁们的共同努力下，在作品展示与思想碰撞中，在"多元化"主题背景烘托中，一定会留下属于这个时代的历史印记。我相信，在主办方北京服装学院的精心筹划下，此次展览一定会推进多元化首饰艺术创作的潮流，实现打造首饰艺术设计全球对话平台的预期愿景，成为一次具有划时代意义的当代首饰艺术盛会。

最后，我谨代表此次活动的指导单位中国工艺美术学会，预祝 2023 北京国际首饰艺术展圆满成功！

才大颖

中国工艺美术学会理事长

SPECH 3

Jewellery holds a unique historical position among the numerous categories of Chinese arts and crafts. It is an independent art category and has intricate connections with the development of other categories. This characteristic endows jewellery with infinite charm of the times. Jewellery, as a decoration for the body, is engraved with long-standing cultural genes and carries human thoughts and emotions towards beauty. Jewellery originated from human worship of nature, synchronized with the origin of human civilization. Through the dialogue between human body decoration and nature, jewellery is endowed with spiritual connotations. The ethnic, cultural, regional, historical, natural, ideological, emotional and other elements contained in it have condensed into a long-standing jewellery design culture and stand out in arts and crafts.

Beijing Institute of Fashion Technology aims to build an internationally renowned university on fashion and has gradually developed into a highland for talent cultivation and fashion innovation in the fashion and cultural creative industries over the years. The Beijing International Jewellery Art Exhibition and the International Jewellery Art Summit Forum have had an influence on the industry. The exhibition building platforms, creating peak dialogues and academic exchanges, and more. Beijing Institute of Fashion Technology has paved a blue channel for industry colleagues to jointly explore the future of jewellery art in the international context. It is not only a result of responsibility but also a sense of mission, demonstrating the pattern and realm of pursuing a first-class university, and implementing the educational purpose of academic service to the country and serving the people's pursuit of a better life.

The theme of this exhibition is " Flourish and Prosper Together ", emphasizing the complementarity and mutual promotion between different cultures, understanding the exquisite realm of jewellery art, and breaking down the barriers of jewellery art, which is a blessing for the arts and crafts industry. It should be a great book in the history of jewellery development. The 2023 Beijing International Jewellery Art Exhibition showcases the power of diversity, aiming to establish a platform for the cultural exchange of diverse jewellery. In this way, we will promote the trend of diversified jewellery art creation and create a broader space for improvement in jewellery art creation. In the future of personalized demand, today's jewellery is no longer just an important artistic medium for improving quality of life, but the attributes and functions of cultural carriers are becoming increasingly prominent. Jewellery constantly breaks through its inherent form, and designers are not only interpreting the unique charm of jewellery, but also not afraid to explore new forms and media, skills and methods. Jewellery design presents a diversified and innovative trend, which is the need of the times. What we need to do is to follow the trend and meet the needs of the times.

Of course, during the exhibition, there will be many topics that require colleagues to openly express their opinions, including but not limited to the following themes: In today's continuous integration of art and fashion, the phenomenon of fashion and commercialization of art does not need to be avoided; the development trends and fashion trends of international jewellery business are issues that jewellery art development must face; Facing the arrival of the wave of artificial intelligence, there may be fundamental changes in the jewellery industry; and topics such as" how sustainable design concepts affect the development of future jewellery education" and "how to carry out jewellery talent cultivation in the context of globalization" are of great concern both inside and outside the industry. I believe that under the careful planning of the organizer, the Beijing Institute of Fashion Technology, this exhibition will promote the trend of diversified jewellery art creation, achieve the expected vision of creating a global dialogue platform for jewellery art design, and become a contemporary jewellery art festival with epoch-making significance.

Finally, on behalf of the guiding unit of this event, China National Arts and Crafts Society, I wish the 2023 Beijing International Jewellery Art Exhibition a complete success!

Cai Daying

Chairman of
China National Arts and Crafts Society

IJCA 设计宣言

人工智能时代，首饰艺术设计更加多元，数字技术、新材料、生物技术等赋予了首饰独特的艺术魅力，越来越多新的形式、媒介、技艺与方法层出不穷。人工智能积极探索首饰艺术多感官互动的形式与关系，在人、社会、环境之间构建一种创新机制。当下，首饰艺术设计教育交织于数字语境中，我们应时刻思考，未来社会需要什么样的人才？在此，面对人工智能、新材料、生物技术等高新技术带来的颠覆性转变，我们共同发布《IJCA 设计宣言 —— AI 时代首饰设计教育发展行动纲领》。

我们倡导

一、教学的组织形式围绕开放、协作和共享展开。它要求我们更加重视开放教育资源和整合渠道，培养学生的智力、社会交往能力和合乎道德的行动能力，以促进学生的有效成长。

二、课程的内容设计围绕生态、跨文化和跨学科学习展开。它要求我们要以更加包容的姿态开展科研和学术活动。培养学生的国际化视野和终身学习能力，以丰富学生获取和创造知识的渠道。

三、教学方法要建立在探究和理解科学之上。科学探究的闭环是观察、质疑、预测、检验、形成理论。这也要求我们要虔诚对待知识的反思、研究、创造以及形成新的教学实践，培养学生的批判性思维，以构建学生的创新能力。

四、教学能力应具备数字世界所需的技能。数字技术蕴含巨大的变革潜能，如何将技术潜力化为现实的路径，才是推动教育改革的关键要素，才能重塑学生、教师、知识和世界之间的关系。这也要求我们要积极拥抱数字化。

今天，探索有效的 AI 时代的教育路径已经迫在眉睫，我们共同倡议，直面人工智能带来的首饰设计教育新变革，整合全球首饰设计教育资源，秉持开放的合作精神，汇聚经验、分享资源，共同致力于推动全球首饰设计教育发展！

IJCA DESIGN DECLARATION

In the era of artificial intelligence, jewellery art and design have become increasingly diverse. Digital technology, new materials, biotechnology, and other factors have bestowed unique artistic charm upon jewellery. New forms, media, techniques, and methods are emerging constantly. Artificial intelligence actively explores multi-sensory interactions and relationships in jewellery art, constructing innovative mechanisms among individuals, society, and the environment. Currently, jewellery art and design education are intertwined within a digital context. We must constantly contemplate what kind of individuals we need to cultivate for the future of society. In this regard, in the face of the disruptive changes brought about by artificial intelligence, new materials, biotechnology, and other high-tech innovations, we unanimously advocate the publication of the "IJCA Design Declaration-Action Plan for the Development of Jewellery Design Education in the AI Era."

We advocate

Firstly, teaching should be organized around openness, collaboration, and sharing. It requires us to pay more attention to the integration of educational resources and channels, nurturing students' intellectual, social, and ethical capabilities to promote effective growth.

Secondly, the curriculum should focus on ecological, cross-cultural, and interdisciplinary learning. It requires us to conduct research and academic activities with a more inclusive attitude. We should cultivate students' international perspectives and lifelong learning capabilities to enrich their channels for acquiring and creating knowledge.

Furthermore, teaching methods should be based on inquiry and understanding of science. The closed loop of scientific inquiry consists of observation, questioning, prediction, testing, and theory formation. This also requires us to sincerely engage in the reflection, research, creation, and development of new teaching practices, fostering students' critical thinking to build their innovation capabilities.

Finally, teaching competence should encompass the skills required in the digital world. Digital technology holds immense transformative potential, and the key to driving educational reform lies in how to turn this technological potential into a reality, reshaping the relationships between students, teachers, knowledge, and the world. This also requires us to actively embrace digitization.

Today, the exploration of effective educational pathways in the AI era is pressing. We jointly propose to confront the new transformation in jewellery design education brought about by artificial intelligence, integrate global jewellery design education resources, uphold an open spirit of collaboration, gather experiences, share resources, and collectively work towards advancing global jewellery design education!

目录
CONTENTS

展览现场
ON THE EXHIBITION

主展区

✦✦ **MAIN EXHIBITION AREA**

教育峰会

31

动态秀

高峰论坛

"未来·域"珠宝首饰设计大赛

✦✦ FUTURE FIELD JEWELLERY DESIGN COMPETITION

国际首饰设计高校联盟大会

✦✦ CONFERENCE OF INTERNATIONAL JEWELLERY COLLEGE ASSOCIATION

参展艺术家
LIST OF ARTISTS

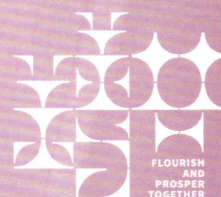

参展艺术家

✦✦ LIST OF ARTISTS

外国参展艺术家
OVERSEA ARTISTS

中国参展艺术家
CHINESE ARTISTS

品牌专区
BRAND ZONE

外国参展艺术家作品

WORKS OF OVERSEA ARTISTS

FLOURISH AND PROSPER TOGETHER

共鸣 1
VIBERATION 1

作者名： Ad Mizrahi（以色列）
类　型： 面饰
材　质： 925 银、珐琅、塑料、缝纫线

Artist: Ad Mizrahi（Israel）
Type: decoration on face
Material: 925 silver, enamel, plastics, sewing thread

跟随 N.11 红点
FOLLWING THE RED DOT N.11

作者名： Adriana Del Duca（意大利）
类　型： 胸针
材　质： 亚克力、钢、925 银、珊瑚膏、硬纸板

Artist: Adriana Del Duca（Italy）
Type: brooch
Material: acrylic, steel, 925 silver, coral paste, pressed cardboard

中点眼镜系列 001
MIDPOINT–EYEWEAR COLLECTION 001

作者名： Ahva Rahminov（以色列）
类　型： 面饰
材　质： 黄铜、925 银、镜片

Artist: Ahva Rahminov（Israel）
Type: decoration on face
Material: brass, 925 silver, lenses

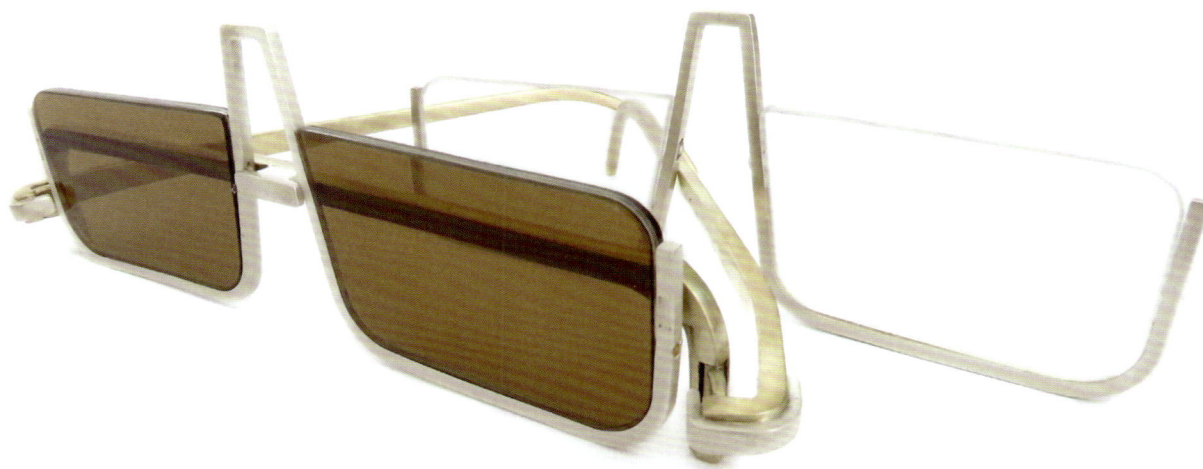

穿拖鞋的牡蛎
OSTRAS EN CHANCLETA

作者名： Aleacion Amarilla（南非）
类　型： 项饰
材　质： 黄铜、铋、橡胶

Artist:　 Aleacion Amarilla（South Africa）
Type:　 necklace
Material: brass, bismuth, rubber

作者名： Aleacion Amarilla（南非）
类　型： 项饰
材　质： 黄铜、铋、橡胶

HW.2
HW.2

作者名： Alice Biolo（意大利）
类　型： 戒指
材　质： 925 银、不锈钢

Artist: Alice Biolo（Italy）
Type: ring
Material: 925 silver, stainless steel

贝都因人
BEDOUIN

作者名： Amal Salim Al-Ismaili（阿曼）
类　型： 手镯
材　质： 皮革、3D 打印尼龙

Artist: Amal Salim Al-Ismaili（Oman）
Type: bracelet
Material: leather, 3D printing nylon

农夫肩上的莫多蒂
MODOTTI IN FARMER'S SHOULDERS

作者名： Ana Lucia Gonzalez Ibanez（墨西哥）
类　型： 项坠、胸针
材　质： 亚克力、纸、925 银、表壳玻璃

Artist:　　Ana Lucia Gonzalez Ibanez（Mexico）
Type:　　pendant, brooch
Material:　acrylic, paper, 925 silver, watch glass

作者名： Ana Lucia Gonzalez Ibanez（墨西哥）
类　型： 项坠、胸针

意志力
WILLPOWER

作者名： Anna Paparella（意大利）
类　型： 项坠
材　质： 炉灰、棉花与塔夫绸面料、照片、
　　　　玻璃珠、USB 数据线、棉纱、丙烯颜料

Artist:　　Anna Paparella（Italy）
Type:　　pendant
Material:　fireplace ash, cotton and taffeta fabric,
　　　　　photographs, glass beads, USB cable,
　　　　　cotton yarn, acrylic paints

焦土
SCORCHED EARTH

作者名： Ariel Lavian（以色列）
类　型： 项圈、胸针
材　质： 紫铜、不锈钢、其他

Artist: Ariel Lavian（Israel）
Type: collar, brooch
Material: copper, stainless steel, others

内与外
INSIDE AND OUTSIDE

作者名： Azin Soltani（伊朗）
类　型： 胸针
材　质： 925 银、钢、砖块、石膏、丙烯颜料

Artist: Azin Soltani（Iran）
Type: brooch
Material: 925 silver, steel, brick, plaster,
acrylic paints

蝴蝶
BUTTERFLY

作者名： Bridget Bailey（英国）
类　型： 装置
材　质： 纸、铁、其他

Artist: Bridget Bailey（England）
Type: installation
Material: paper, iron, others

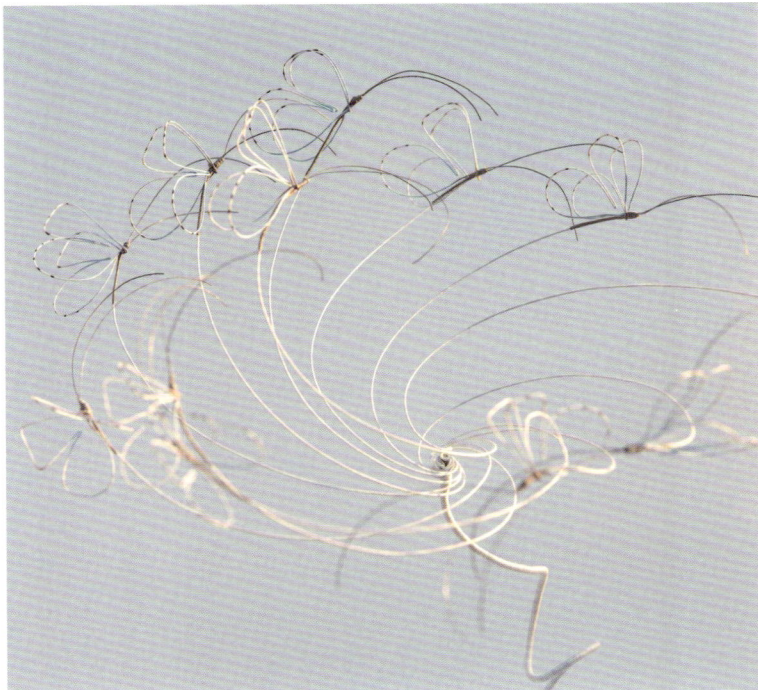

走到一起
COME TOGETHER

作者名： Cecilia Lopez Bravo（美国）
类　型： 戒指、项坠
材　质： 玻璃、黄铜、不锈钢、
　　　　24K 金、纯银

Artist: Cecilia Lopez Bravo（USA）
Type: ring, pendant
Material: glass, brass, stainless steel,
　　　　24K gold, pure silve

治愈性叙事记忆
HEALING NARRATIVE MEMORY

作者名： Chandra Ngomane（南非）
类　型： 手镯
材　质： 925 银、陶瓷、玻璃

Artist: Chandra Ngomane（South Africa）
Type: bracelet
Material: 925 silver, ceramic, glass

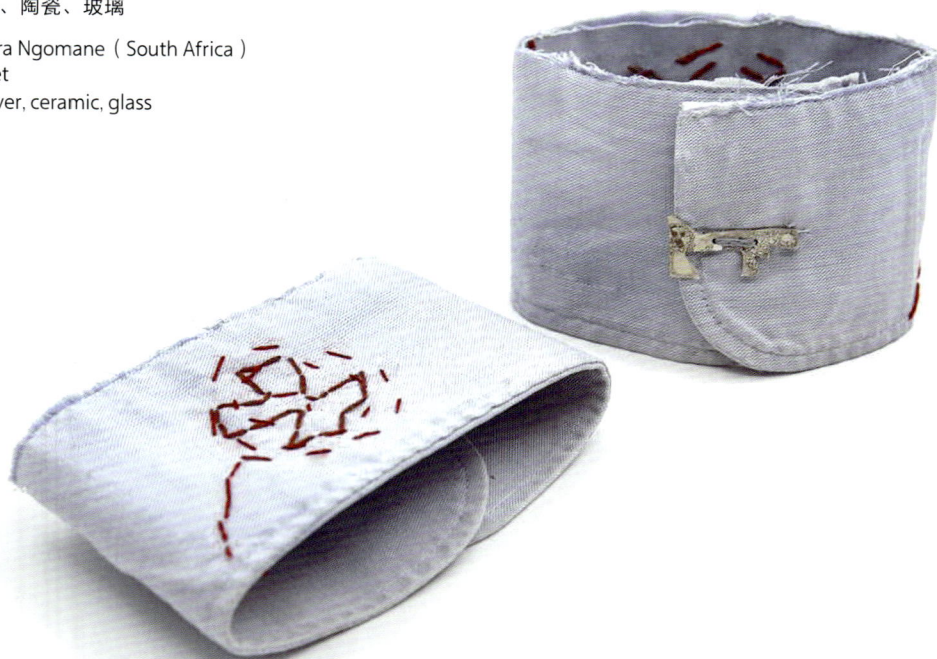

重温
REREADINGS

作者名： Chiara Trentin（意大利）
类　型： 项饰
材　质： 塑料、古旧纽扣、
　　　　 古旧锁扣、打蜡棉线

Artist: Chiara Trentin（Italy）
Type: necklace
Material: plastics, vintage buttons, vintage buckle,
waxed cotton thread

震感动态戒指
THE VIBRONIC KINETIC RING

作者名： Claudio Pino（加拿大）
类　型： 戒指
材　质： 银、18K 金、19K 金、
　　　　 黑玛瑙、欧泊、石榴石

Artist: Claudio Pino（Canada）
Type: ring
Material: silver, 18K gold, 19K gold,
black onyx, opal, garnet

蓝色、棕色与白色
BLUE, BROWN AND WHITE

作者名： Daniel von Weinberger（比利时）
类　型： 项饰
材　质： 塑料

Artist:　　Daniel von Weinberger（Belgium）
Type:　　 necklace
Material:　plastic

现在来见我
NOW YOU COME ME

作者名： Dodo Paruznik（奥地利）
类　型： 胸针
材　质： 树脂、钢、塑料、
　　　　 黄铜、纸、包装材料、金属链

Artist: Dodo Paruznik（Austria）
Type: brooch
Material: resin, steel, plastics, brass, paper,
packaging materials, metal chain

不平等 NO.1
INEQUALITY NO.1

作者名： Eden Danieli（以色列）
类　型： 胸针、项饰
材　质： 银、钢、施华洛世奇水晶

Artist: Eden Danieli（Israel）
Type: brooch, necklace
Material: silver, steel, swarovski stone

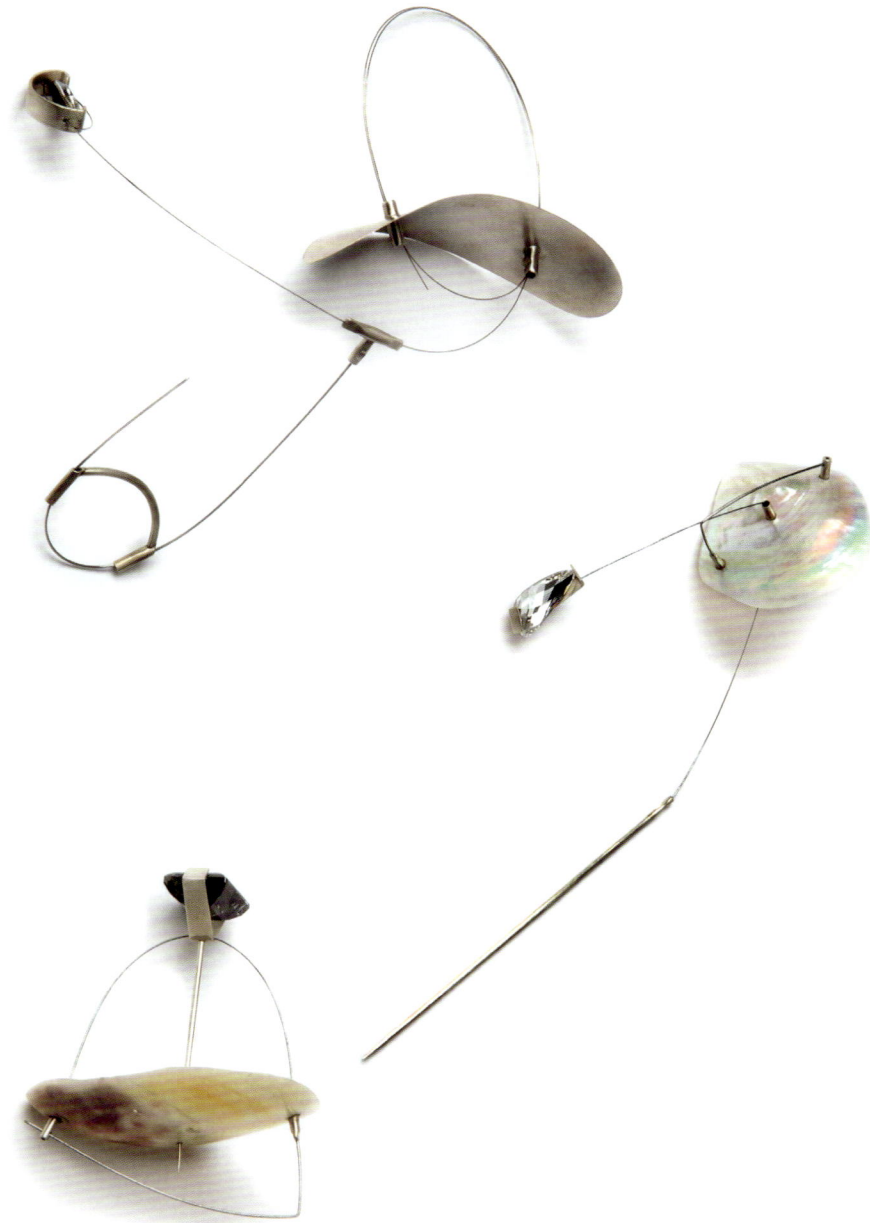

义肢 #1
PROSTHETIC #1

作者名： Gabriela Ramirez Michel（墨西哥）
类　型： 手饰
材　质： 陶瓷、珐琅

Artist: Gabriela Ramirez Michel（Mexico）
Type: decoration on hand
Material: ceramic, enamel

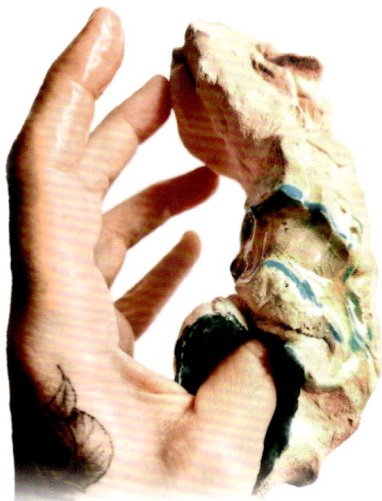

没有看到花纹的织物
NOT SEEING THE FABRIC FOR THE TREAD

作者名： Gussie van der Merwe（南非）
类　型： 项坠、耳饰
材　质： 925 银、铜、棉绳

Artist: 　Gussie van der Merwe（South Africa）
Type: 　　pendant, earrings
Material: 　925 silver, copper, cotton cord

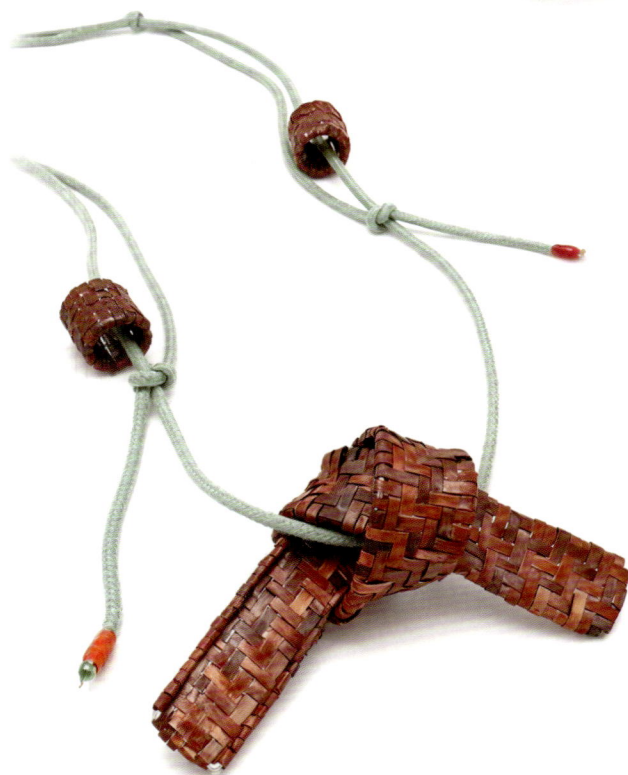

延伸
EXTENSION

作者名： Hasan Kurd（巴勒斯坦）
类　型： 手饰
材　质： 大理石

Artist:　Hasan Kurd（Palestine）
Type:　decoration on hand
Material:　marble

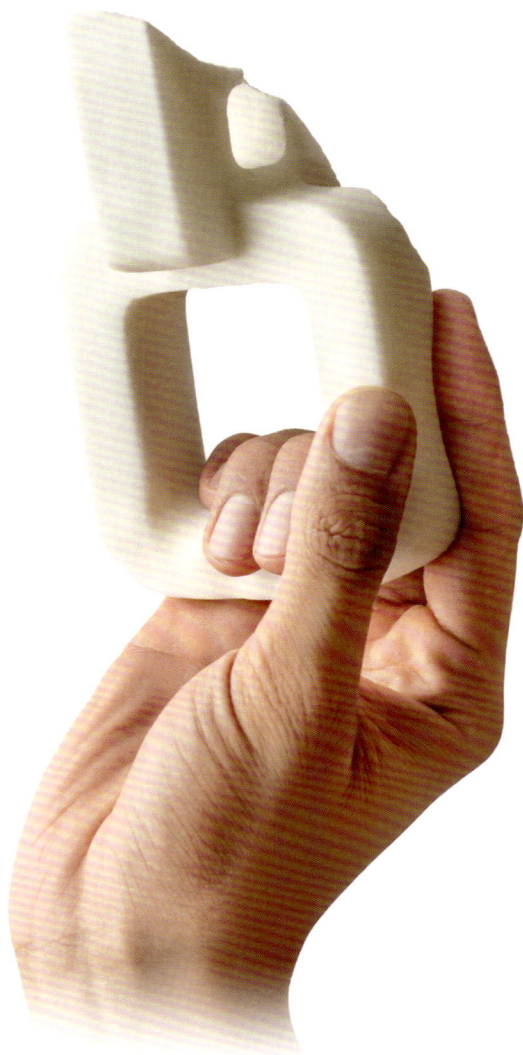

穆萨·帕克壁画发夹
MUSA PAK FRESCO

作者名： Irsa Khan（巴基斯坦）
类　型： 项坠、发饰
材　质： 银、骆驼骨、21K 金、珐琅

Artist: Irsa Khan（Pakistan）
Type: pendant, hair ornament
Material: silver, camel bone, 21K gold, enamel

作者名： Irsa Khan（巴基斯坦）

绣花珐琅
EMBROIDERED ENAMEL

作者名： Jane Moore（英国）
类　型： 胸针
材　质： 纯银、珐琅

Artist: Jane Moore（England）
Type: brooch
Material: pure silver, enamel

重生
REBORN LIFE

作者名： Jean Marc WASZACK（法国）
类　型： 项链、胸针、发饰
材　质： 钛、珍珠、鹿骨

Artist: Jean Marc WASZACK（France）
Type: necklace, brooch, hair ornament
Material: titanium, pearl, fawn skull

夏娃之泪
TEARS OF EVE

作者名： Jivan Astfalck（德国）
类　型： 胸针
材　质： 24K 金、陶瓷、925 银、其他

Artist: Jivan Astfalck（Germang）
Type: brooch
Material: 24K gold, ceramic, 925 silver, others

彩色的充气戒指
PUFFED UP RING IN COLOR

作者名： Kim Buck（丹麦）
类　型： 戒指
材　质： 纯银、珐琅涂层

Artist: Kim Buck（Denmark）
Type: ring
Material: pure silver, enamel coated

梦幻十二
DREAM XII

作者名： Kim Jiseo（韩国）
类　型： 项圈、胸针
材　质： 24K 金、大漆、金箔、木材、
　　　　 紫铜、纸、黄铜、银

Artist:　Kim Jiseo（South Korea）
Type:　necklace, brooch
Material:　24K gold, lacquer, gold leaf, wood,
　　　　 copper, paper, brass, silver

流星
SHOOTING STAR

作者名： Kim Jiyoung（韩国）
类　型： 胸针
材　质： 不锈钢、其他

Artist: Kim Jiyoung（South Korea）
Type: brooch
Material: stainless steel, others

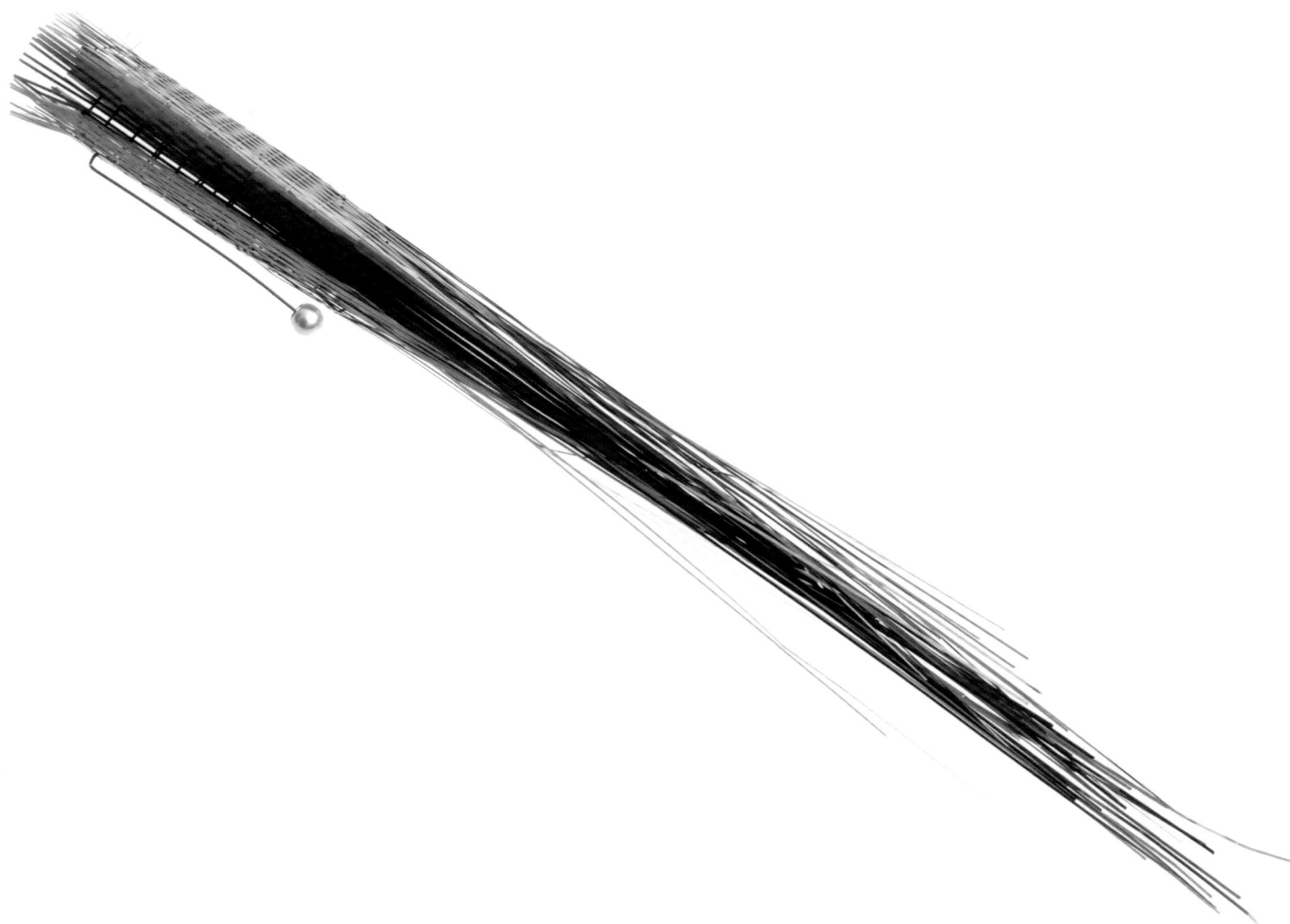

扩散
SPREAD

作者名： Kim Sunae（韩国）
类　型： 手镯
材　质： 紫铜、银、其他

Artist:　　Kim Sunae（South Korea）
Type:　　　bracelet
Material:　copper, silver, others

失落花园的故事：第一章
THE STORY OF LOST GARDEN : CHAPTER 1

作者名： Kimia Parang（伊朗）
类 型： 戒指
材 质： 银、925 银、红宝石、其他

Artist: Kimia Parang（Iran）
Type: ring
Material: silver, 925 silver, ruby, others

鼻垫系列 #03
NOSE PADS SERIES #03

作者名： Kong Saerom（韩国）
类　型： 项坠
材　质： 银、硅胶

Artist:　　Kong Saerom（South Korea）
Type:　　necklace
Material:　silver, silicon

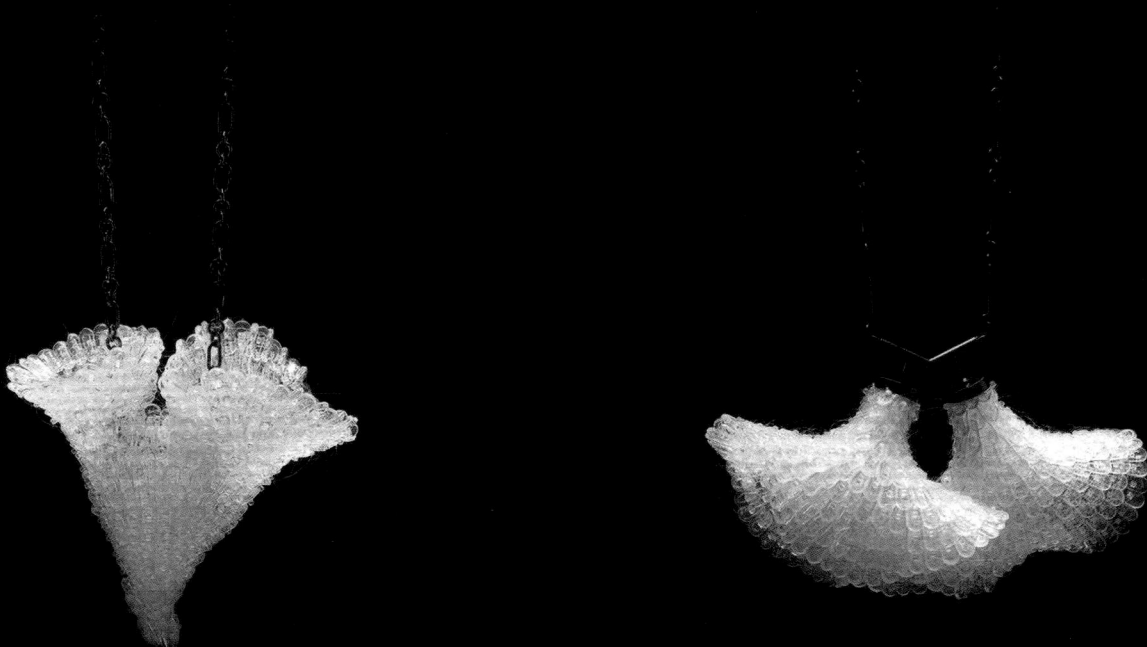

鼻垫系列 #03
NOSE PADS SERIES #03

作者名： Kong Saerom（韩国）
类　型： 项坠
材　质： 银、硅胶

胶囊 1 号
CAPSULE 1

作者名： Laura Forte（意大利）
类　型： 戒指
材　质： 塑料、铝

Artist: Laura Forte（Italy）
Type: ring
Material: plastics, aluminium

作者名： Laura Forte（意大利）
类　型： 戒指
材　质： 塑料、铝

填充空隙的装置（14）
DEVICE FOR FILLING A VOID (14)

作者名： Lauren Kalman（美国）
类　型： **身体装饰**
材　质： **紫铜、黄铜、紫铜镀金**

Artist: Lauren Kalman（USA）
Type: body adornment
Material: copper, brass, gold-plated copper

锯切和压制 4
SAWED & PRESSED 4

作者名： Lee Yeonmi（韩国）
类　型： 胸针
材　质： 氧化 925 银

Artist: 　Lee Yeonmi（South Korea）
Type: 　brooch
Material: 　oxidized 925 silver

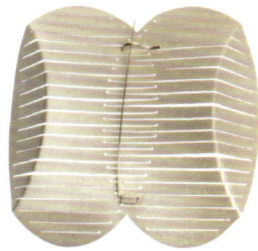

超越
BEYOND

作者名： Margit Hart（奥地利）
类　型： 胸针
材　质： 铝、银、玛瑙

Artist:　　Margit Hart（Austria）
Type:　　 brooch
Material:　aluminum, silver, agate

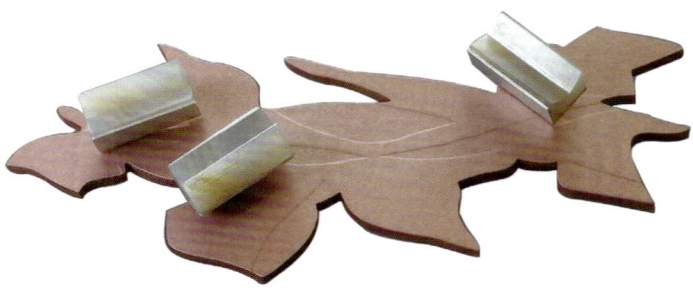

无题
UNTITLED

作者名： Maria Rosa Franzin（意大利）
类　型： 项饰、胸针
材　质： 链条、银、钢、珊瑚、塑料纸

Artist:　　Maria Rosa Franzin（Italy）
Type:　　necklace, brooch
Material:　chain, silver, steel, coral, plastic paper

解构
DECONSTRUCTION

作者名： Maya Shochat（以色列）
类　型： 戒指
材　质： 925 银、氧化锆

Artist: Maya Shochat（Israel）
Type: ring
Material: 925 silver, zirconia

重画正方形
REDRAW SQUARE

作者名： Min Bogki（韩国）
类　型： 胸针
材　质： 金箔、铝、亚克力

Artist:　Min Bogki（South Korea）
Type:　brooch
Material:　gold leaf, aluminum, acrylic

致　谢： 本项创作获得由韩国政府（MSIT）
（编号 RS-2023-00219429）资助的
韩国国家研究基金会（NRF）的支持

Acknowledgement:
This work was supported by
the National Research Foundation of Korea (NRF)
grant funded by the Korea government (MSIT)
(No. RS-2023-00219429)

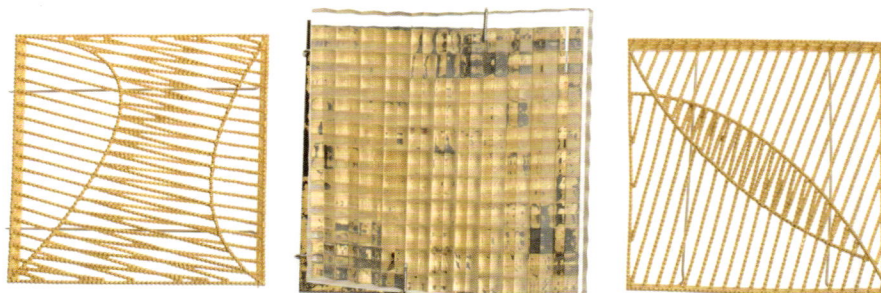

化石 01
FOSSIL 01

作者名： Omer Adar（以色列）
类　型： 胸针
材　质： 黄铜、金属编织线

Artist: Omer Adar（Israel）
Type: brooch
Material: brass, metal braiding thread

一起来玩捉迷藏
LET'S PLAY HIDE AND SEEK

作者名： Poly Iglesias（阿根廷）
类　型： 胸针
材　质： 不锈钢、珐琅、紫铜、
　　　　 青金石、青铜、金属丝

Artist: Poly Iglesias（Argentina）
Type: brooch
Material: stainless steel, enamel,
copper, lapis lazuli, bronze, wire

卡梅奥 1 号
CAMEO I

作者名： Sara Gackowska（波兰）
类　型： 胸针
材　质： 纸、钢、摄影纸、
　　　　 胶水、羊驼毛

Artist: Sara Gackowska（Poland）
Type: brooch
Material: paper, steel, photography paper,
glue, alpaca

冬芽
WINTER BUD

作者名： Sara Shahak（以色列）
类　型： 胸针
材　质： 不锈钢、铁、玻璃涂料

Artist: Sara Shahak（Israel）
Type: brooch
Material: stainless steel, iron, glass paint

倾斜和拖动 1 —— 扁平化
TILT & DRAG 1 — FLATTENED

作者名： Shin JaKyung（韩国）
类　型： 胸针
材　质： 不锈钢、18K 金

Artist:　　Shin JaKyung（South Korea）
Type:　　brooch
Material:　stainless steel, 18K gold

作者名： Shin JaKyung（韩国）
类　型： 胸针
材　质： 不锈钢、18K 金

水中的碎片
FRAGMENTS IN WATER

作者名： Song Yookyung（韩国）
类　型： 胸针
材　质： 925 银、树脂、亚克力、镓

Artist:　　Song Yookyung（South Korea）
Type:　　 brooch
Material:　925 silver, resin, acrylic, gallium

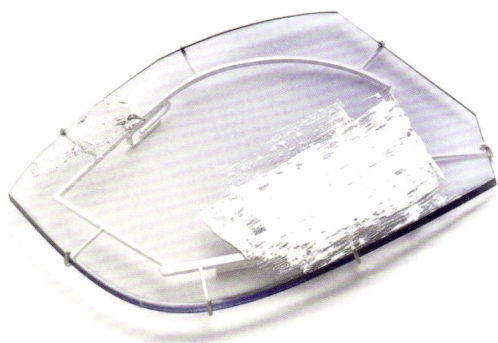

奖杯头
TROPHY HEAD

作者名： Song Yuxin（加拿大）
类　型： 胸针
材　质： 紫铜、珐琅、陶瓷、
　　　　 不锈钢、银、其他

Artist: Song Yuxin（Canada）
Type: brooch
Material: copper, enamel, ceramic,
stainless steel, silver, others

作者名： Song Yuxin（加拿大）
类　型： 胸针

贝利库斯
BERICUS

作者名： Stefano Rossi（意大利）
类　型： 胸针
材　质： 木纹金、银、钢

Artist: Stefano Rossi（Italy）
Type: brooch
Material: mokume gane, silver, steel

无题
UNTITLED

作者名： Stephie Morawetz（奥地利）
类　型： 胸针
材　质： 不锈钢、树脂、亚克力、塑料、人造石

Artist: Stephie Morawetz（Austria）
Type: brooch
Material: stainless steel, resin, acrylic,
　　　　　plastics, artificial stone

峡系列
GORGE SERIES

作者名： 王春刚（美国）
类　型： 吊坠
材　质： 珍珠、钻石、蓝宝石、
　　　　 珐琅、银

Artist:　　Wang Chungang（USA）
Type:　　　pendant
Material:　pearl, diamond, sapphire,
　　　　　 enamel, silver

切屑 — 切掉
CHIP – CHIP AWAY

作者名： Zofia Skoroszewska（波兰）
类　型： 装置
材　质： 沙子、实验室蓝宝石、纯银

Artist:　　Zofia Skoroszewska（Poland）
Type:　　　installation
Material:　sand, laboratory sapphire, pure silver

中国参展
艺术家作品
WORKS OF
CHINESE ARTISTS

渔的捆绑
BINDING OF FISHING

作者名：　白冷泠（中国）
类　型：　胸针
材　质：　树脂、塑料、黄铜、
　　　　　玻璃、废弃电线、渔线

Artist:　　Bai Lengling（China）
Type:　　 brooch
Material:　resin, plastic, brass, glass,
　　　　　abandoned wire, fishing line

轻如鸿毛之一
NO.1 BE AS LIGHT AS A FEATHER

作者名： 白晓宇（中国）
类　型： 戒指、胸针、项饰
材　质： 14K 银丝、蚕丝、紫铜、其他

Artist:　　Bai Xiaoyu（China）
Type:　　 ring, brooch, necklace
Material: 14K silver wire, silk, copper, others

轻如鸿毛之一
NO.1 BE AS LIGHT AS A FEATHER

作者名： 白晓宇（中国）
类　型： 戒指、胸针、项饰

《洞·见》2023 系列
INSIGHT 2023 SERIES

作者名：　白一宏（中国）
类　型：　胸针
材　质：　925 银、蓝宝石、银、丝线、
　　　　　蓝宝石、丝绸、其他

Artist:　　Bai Yihong（China）
Type:　　　brooch
Material:　925 silver, sapphire, silver, silk thread,
　　　　　sapphire, silk cloth, others

纪念
REMEMBER

作者名： **拜月姣（中国）**
类　型： **项饰**
材　质： **925 银、纸、亚克力、轻黏土**

Artist: **Bai Yuejiao（China）**
Type: **necklace**
Material: **925 silver, paper, acrylic, light clay**

织路之谱
THE SPECTRUM OF WEAVING

作者名： 鲍蕊（中国）
类　型： 项饰、胸针
材　质： 925 银、珐琅、紫铜

Artist: Bao Rui（China）
Type: necklace, brooch
Material: 925 silver, enamel, copper

回忆的载体
MEMORABLE OBJECTS

作者名： 蔡倩翘（中国香港）
类　型： 胸针
材　质： 黄铜、木材

Artist:　　Choi Sin Kiu（Hong Kong, China）
Type:　　 brooch
Material: brass, wood

是你，是我，也是他（她）
YOU ARE ME, AND I AM YOU,
AS WELL AS HIM AND HER

作者名： 蔡锐龙（中国）
类 型： 胸针
材 质： 925 银、银镀金

Artist: Cai Ruilong（China）
Type: brooch
Material: 925 silver, gold-plated silver

大同·小异系列
SIMILARITY · MINOR DIFFERENCES SERIES

作者名： 陈彬雨（中国）
类 型： 戒指
材 质： 锆石、925 银、银镀金、其他

Artist: Chen Binyu（China）
Type: ring
Material: zircon, 925 silver, gold-plated silver, others

生命之轻
LIGHTNRESS OF BEING

作者名： 陈乐怡（中国）
类　型： 项饰、胸针
材　质： 糙米、红香米、黑米、
　　　　　皮革、蜡线、其他

Artist:　　Chen Leyi（China）
Type:　　necklace, brooch
Material:　brown rice, red fragrant rice, black rice,
　　　　　leather, wax string, others

八十亿分之一
ONE IN EIGHT BILLION

作者名： **陈镕岚（中国）**
类　型： **戒指**
材　质： **925 银、钢丝、铜版纸标签**

Artist:　　Chen Ronglan（China）
Type:　　ring
Material:　925 silver, steel wire, glossy paper label

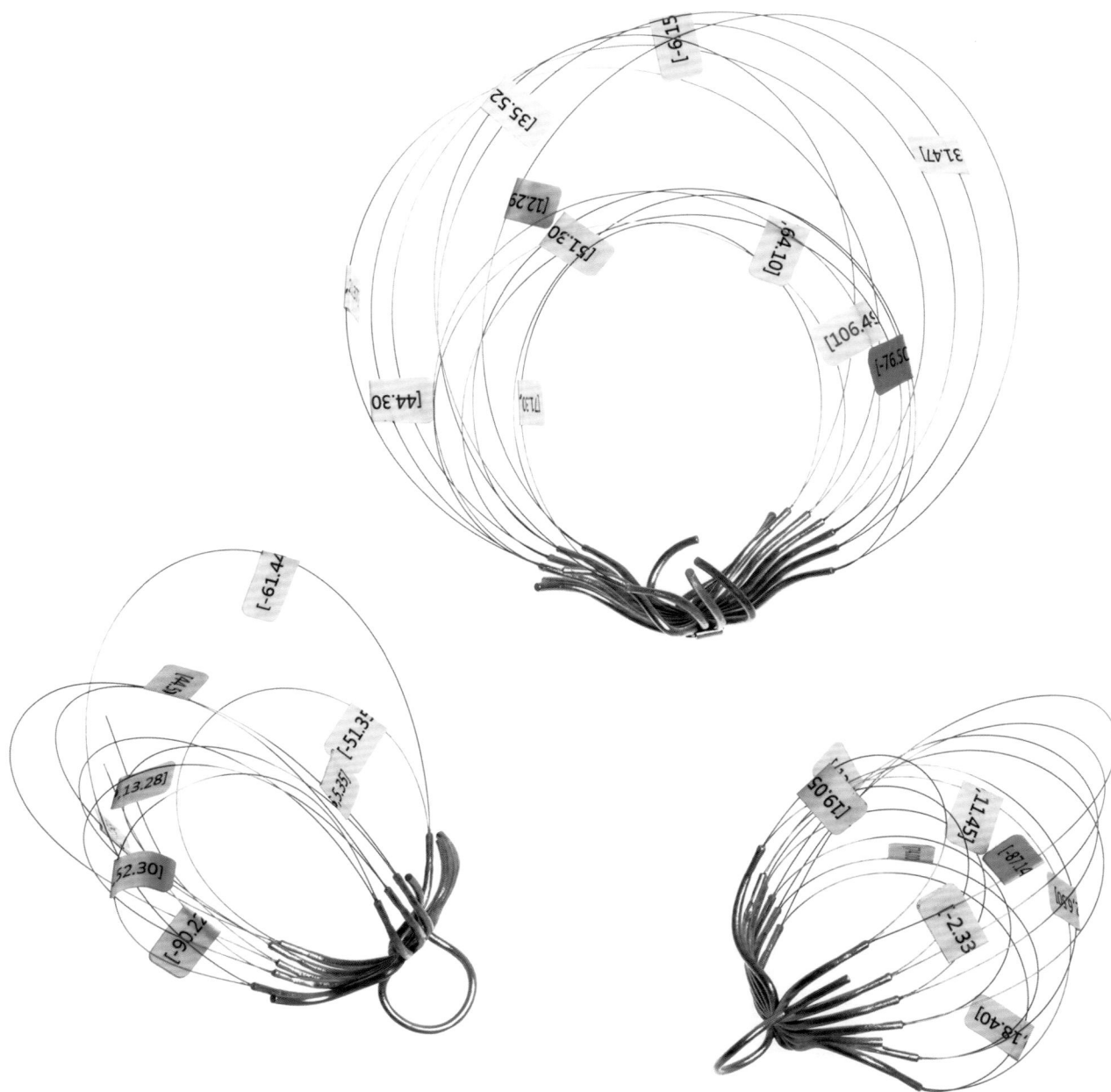

中国醒狮
CHINESE LION

作者名：　陈启彬（中国）
类　型：　戒指、耳环、吊坠、摆件
材　质：　紫铜、莫桑石、
　　　　　镀铂金、朋克低温珐琅

Artist:　　Chen Qibin（China）
Type:　　　ring, earrings, pendant, decoration
Material:　copper, moissanite, platinum-plated,
　　　　　　punk low-temperature enamel

瓷青的爱
THE LOVE OF PORCELAIN BLUE

作者名： 陈卓（中国）
类　型： 胸针
材　质： 925 银、珍珠、珐琅、皮革、其他

Artist:　　Chen Zhuo（China）
Type:　　 brooch
Material:　925 silver, pearl, enamel, leather, others

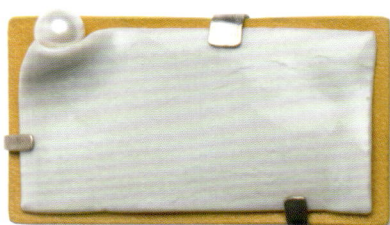

我将宇宙随身携带
I CARRY THE UNIVERSE WITH ME

作者名： 丛聪（中国）
类 型： 耳饰、项饰
材 质： 黄铜镀金、珍珠、其他

Artist: Cong Cong（China）
Type: earrings, necklace
Material: gold-plated copper, pearl, others

作者名： 丛聪（中国）

观想之一
VISUALIZATION · NO.1

作者名： 崔金玉（中国）
类 型： 戒指
材 质： 925 银

Artist: Cui Jinyu（China）
Type: ring
Material: 925 silver

图腾系列 1
TOTEM 1

作者名： 戴翔（中国）
类　型： 项饰、胸针
材　质： 木材、皮绳、丙烯颜料

Artist: Dai Xiang（China）
Type: necklace, brooch
Material: wood, leather cord, acrylic paints

鱼
FISH

作者名： 邓婧怡（中国）
类　型： 胸针
材　质： 钛、通天玉

Artist: Deng Jingyi（China）
Type: brooch
Material: titanium, heavenly jade

太极
THE GREAT ULTIMATE

作者名： 丁晓飞（中国）
类　型： 胸针
材　质： 钛、钛金

Artist: Ding Xiaofei（China）
Type: brooch
Material: titanium, titanium gold

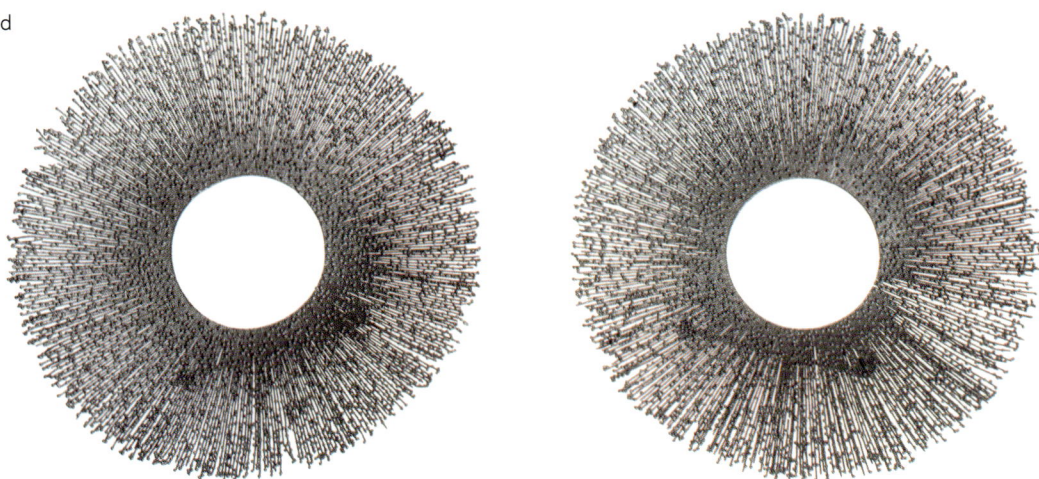

衔花过江湖
CARRING FLOWER ACROSS THE JIANGHU

作者名： 豆肖楠（中国）
类　型： 胸针
材　质： 红宝石、玛瑙、银镀金

Artist:　　Dou Xiaonan（China）
Type:　　 brooch
Material:　ruby, agate, gold-plated silver

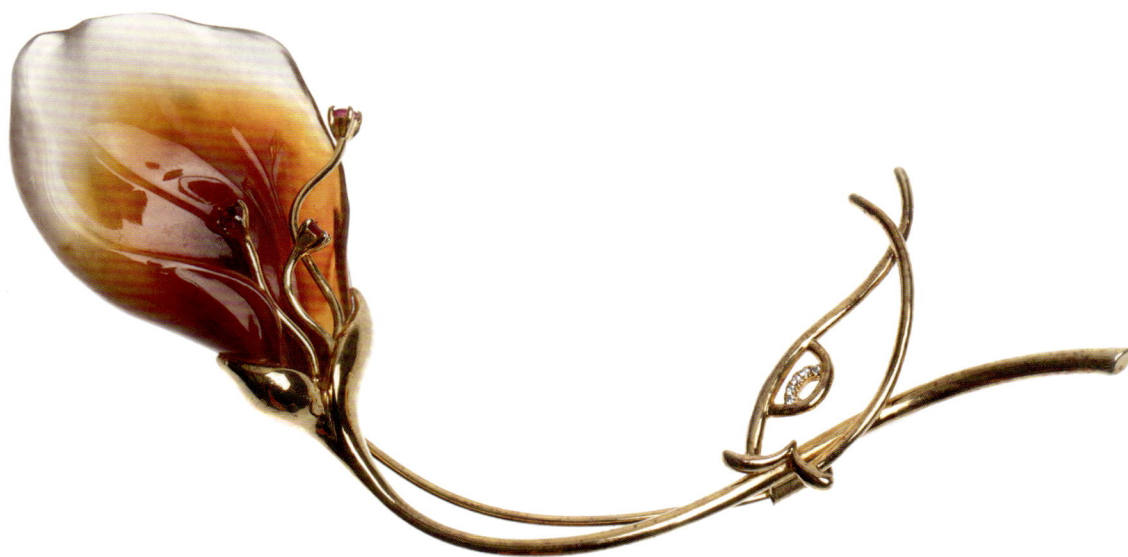

荣华富贵
WEALTH & RANK

作者名： 杜建毅（中国）
类　型： 项链
材　质： 18K 金、珊瑚、钻石

Artist: Du Jianyi（China）
Type: necklace
Material: 18K gold, coral, diamond

永恒的时光轮回
ETERNAL CYCLE OF TIME

作者名： 段润发、洪书瑶（中国）
类 型： 项饰、手镯
材 质： 塑料、ABS 塑料

Artist: Duan Runfa, Hong Shuyao（China）
Type: necklace, bracelet
Material: plastic, ABS plastic

作者名： 段润发、洪书瑶（中国）
类 型： 项饰、手镯
材 质： 塑料、ABS 塑料

梦回云冈系列之八
DREAM BACK TO YUNGANG GROTTOES – 8

作者名： **段燕丽（中国）**
类　型： **胸针**
材　质： **纯银、珐琅、白玉髓**

Artist:　　Duan Yanli（China）
Type:　　brooch
Material:　pure silver, enamel, white chalcedony

吉至
AUSPICIOUS ARRIVAL

作者名： 房文红（中国）
类　型： 耳饰
材　质： 18K 金、祖母绿、红宝石、
　　　　蓝宝石、沙弗莱、尖晶石、钻石

Artist:　　Fang Wenhong（China）
Type:　　earrings
Material:　18K gold, emerald, ruby, sapphire,
　　　　　tsavorite, spinel, diamond

迁徙，候鸟
THE MIGRATION, MIGRATORY BIRD

作者名： 冯婉婷（中国香港）
类　型： 胸针
材　质： 聚乳酸、镜子、丙烯酸、铝

Artist: 　Fung Autumn（Hong Kong, China）
Type: 　brooch
Material: 　polylactic acid, mirrors, acrylic, aluminium

梦想的种子系列二
THE SEED OF DREAMS SERIES 2

作者名： 冯雪晶、赵慧颖（中国）
类　型： 胸针
材　质： 纯银、珐琅

Artist:　　Feng Xuejing, Zhao Huiying（China）
Type:　　 brooch
Material:　pure silver, enamel

幻星流云
ENCHANTED STARRY CLOUDS

作者名： 冯艺佳（中国）
类　型： 项圈、胸针、戒指
材　质： 纯银、树脂、丙烯

Artist:　　Feng Yijia（China）
Type:　　 necklace, brooch, ring
Material:　pure silver, resin, propylene

贯·注
GUAN · ZHU

作者名： 付少雄（中国）
类 型： 胸针
材 质： 925 银、金

Artist: Fu Shaoxiong（China）
Type: brooch
Material: 925 silver, gold

痴竹拜石 II
OBSESSED WITH BAMBOO AND TAIHU STONE II

作者名： 傅永和（中国）
类　型： 项坠、胸针
材　质： 紫铜、珐琅

Artist: 　Fu Yonghe（China）
Type: 　　pendant , brooch
Material: 　cooper, enamel

山 · 海
MOUNTAIN · SEA

作者名： 高珊（中国）
类　型： 项饰
材　质： 925 银、玻璃

Artist: Gao Shan（China）
Type: necklace
Material: 925 silver, glass

古韵新声
ANCIENT RHYTHMS, MODERN ECHOES

作者名： 高伟（中国）
类 型： 项链
材 质： 925 银、彩色玛瑙

Artist: Gao Wei（China）
Type: necklace
Material: 925 silver, colored agate

垃圾计划 2023 —— 生长
GARBAGE PLAN 2023 — GROW

作者名： 巩志伟（中国）
类　型： 胸针
材　质： 纸、纯银、珐琅、
　　　　紫铜、黄铜、钢

Artist:　　Gong Zhiwei（China）
Type:　　 brooch
Material:　paper, pure silver, enamel,
　　　　　copper, brass, steel

青 · 金
BLUE · GOLD

作者名：　**古丽米拉 · 艾尼（中国）**
类　型：　**项饰**
材　质：　**黄铜、青金石、其他**

Artist:　　Gulmira · Aine（China）
Type:　　　necklace
Material:　brass, lapis lazuli, others

蛙声一片
SOUNDS OF FROGS

作者名: 谷明（中国）
类　型: 项饰
材　质: 纯银、木材、钧瓷、其他

Artist: Gu Ming（China）
Type: necklace
Material: pure silver, wood, Jun porcelain, others

我和我的丑系列 2
ME AND MY UGLINESS SERIES

作者名： 管管（中国）
类 型： 装置
材 质： 金丝

Artist: Guan Guan（China）
Type: installation
Material: gold thread

"蟹" 谢
"CRAB" THANK

作者名： 郭涛（中国）
类　型： 胸针
材　质： 纯银、黄铜

Artist: Guo Tao（China）
Type: brooch
Material: pure silver, brass

22 世纪 –F2
22ND CENTURY – F2

作者名： 郭伟韬（中国）
类　型： 耳饰
材　质： PLA 打印材料、925 银、大漆、其他

Artist: Guo Weitao（China）
Type: earrings
Material: PLA printing materials, 925 silver, lacquer, others

五兽丰登
BEAST JOY

作者名： 韩欣然（中国）
类　型： 项坠、耳饰
材　质： 925 银

Artist: Han Xinran（China）
Type: pendant, earrings
Material: 925 silver

小熊座
LITTLE DIPPER

作者名： **韩雨蒙（中国）**
类　型： **胸针**
材　质： **6061 铝合金、925 银**

Artist:　　 Han Yumeng（China）
Type:　　　 brooch
Material:　 6061 aluminium alloy, 925 silver

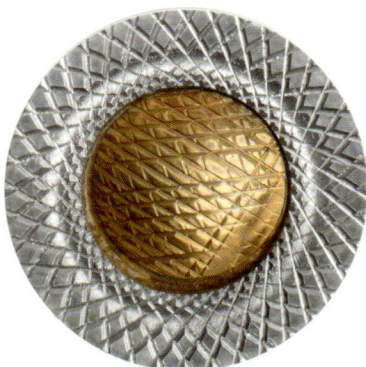

花朵
FLOWER

作者名： **何彦欣（中国）**
类　型： **胸针**
材　质： **925 银、紫铜**

Artist: He Yanxin（China）
Type: brooch
Material: 925 silver, copper

落日
SUNSET

作者名： 侯安丽（中国）
类　型： 胸针
材　质： 纯银、珍珠、塑料、
　　　　 银箔、人造宝石

Artist: Hou Anli（China）
Type: brooch
Material: pure silver, pearl, plastic,
silver foil, artificial stone

疏影系列 – 1
SCATTERED SHADOWS – 1

作者名： 胡世法（中国）
类　型： 胸针
材　质： 纯银、紫铜、钢、铝、
　　　　 玻璃珠、丙烯、其他

Artist:　　Hu Shifa（China）
Type:　　　brooch
Material:　pure silver, copper, steel, aluminum,
　　　　　 glass beads, propylene, others

数码人 2#
DIGITAL MAN 2#

作者名： 胡俊（中国）
类 型： 胸针
材 质： 尼龙、银镀金、镜面、钢丝

Artist: Hu Jun（China）
Type: brooch
Material: nylon, gold plated silver, mirror, steel wire

塑料悲剧
PLASTIC TRAGEDY

作者名： 黄可欣（中国）
类　型： 项坠、胸针
材　质： 黄铜、塑料

Artist:　　Huang Kexin（China）
Type:　　 pendant, brooch
Material:　brass, plastic

无双
TWO POINTS IN ONE

作者名： **黄琳（中国）**
类　型： **胸针**
材　质： **925 银、陶瓷**

Artist: **Huang Lin（China）**
Type: brooch
Material: 925 silver, ceramic

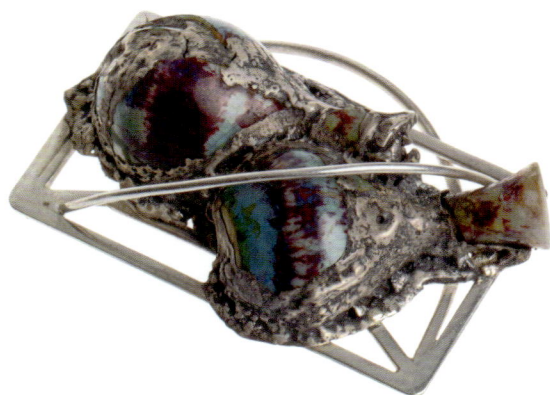

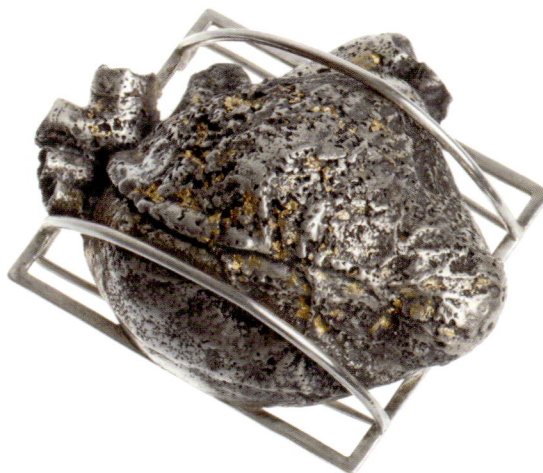

作者名： **黄琳（中国）**

瓶 · 戒 · 链 – 5
VASE · RING · CHAIN – 5

作者名： 霍霓（中国）
类 型： 项链、戒指
材 质： 925 银、树脂、锆石

Artist: Huo Ni（China）
Type: necklace, ring
Material: 925 silver, resin, zircon

苏州园林
SUZHOU GARDEN

作者名： 吉毓熹（中国）
类 型： 胸针
材 质： 925 银、苏绣、真丝、其他

Artist: Ji Yuxi（China）
Type: brooch
Material: 925 silver, suzhou embroidery,
silk, others

不朽
THE IMMORTAL SPIRIT

作者名： 冀婕（中国）
类　型： 胸饰
材　质： 黄铜、铜镀 18K 金、铜镀铑

Artist:　　Ji Jie（China）
Type:　　　brooch
Material:　brass, 18K gold-planted copper,
　　　　　 rhodium-planted copper

绽 –1
BLOOMING – 1

作者名： 姜倩（中国）
类　型： 项圈
材　质： 纯银

Artist: 　Jiang Qian（China）
Type: 　 necklace
Material:　pure silver

善道
GOOD WAY

作者名： 金翠玲（中国）
类　型： 胸针
材　质： 925 银、纸

Artist:　　Jin Cuiling（China）
Type:　　　brooch
Material:　 925 silver, paper

致敬陈世英
SALUTE TO WALLACE CHAN

作者名： **金磊（中国）**
类　型： **胸针**
材　质： **钛、蓝宝石、红宝石、
　　　　钻石、黄水晶**

Artist:　　Jin Lei（China）
Type:　　brooch
Material:　titanium, sapphire, ruby,
　　　　　diamond, citrine

作者名： **金磊（中国）**

蝶舞
BUTTERFLY DANCE

作者名： **李楠（中国）**
类　型： **胸针**
材　质： **18K 金、树脂、珍珠漆**

Artist:　　**Li Nan（China）**
Type:　　　brooch
Material:　**18K gold, resin, nacrolacquer**

蝴蝶结
BOW

作者名： **李骞（中国）**
类　型： **项饰**
材　质： **925 银**

Artist: **Li Qian（China）**
Type: necklace
Material: 925 silver

二维解构 —— 相拥
2ND DESCONSTRUCTION —
EMBRACE EACH OTHER

作者名： **李泉颖（中国）**
类　型： **耳饰**
材　质： **970 银**

Artist: **Li Quanying（China）**
Type: **earrings**
Material: **970 silver**

江南相思引
MISSING FAIR SOUTH

作者名： 李桑（中国）
类 型： 项饰
材 质： 纯银

Artist: Li Sang（China）
Type: necklace
Material: pure silver

作者名： 李桑（中国）

植物的灵魂 —— 捡拾落叶
SOULS OF PLANTS —
PICKING UP LEAVES

作者名： 李苑源（中国）
类　型： 胸针
材　质： 纯银、不锈钢、尼龙网、
　　　　 蚕丝线、其他

Artist:　　Li Yuanyuan（China）
Type:　　 brooch
Material:　pure silver, stainless steel,
　　　　　nylon net, silk thread, others

孟夏
THE FIRST MONTH OF SUMMER

作者名： 梁大钊（中国）
类　型： 胸针
材　质： 钛金、钻石、蓝宝石、水晶

Artist:　　Liang Dazhao（China）
Type:　　brooch
Material:　titanium, diamond, sapphire, crystal

简
SIMPLE

作者名： 梁燕观（中国）
类　型： 耳饰
材　质： 钛金、铝、托帕石、其他

Artist: Liang Yanguan（China）
Type: earrings
Material: titanium, aluminum, topaz, others

荆棘之心
HEART OF THORNS

作者名： 梁由之（中国）
类　型： 项饰、胸针
材　质： 925 银、大漆、树脂、锆石、螺钿

Artist:　Liang Youzhi（China）
Type:　necklace, brooch
Material:　925 silver, lacquer, resin,
　　　　　zircon, mother-of-pearl

盛势绽放
FULL OF LIFE

作者名： 林弘裕（中国）
类　型： 胸针
材　质： 天然红珊瑚、天然红宝石、天然钻石、
　　　　 天然翡翠、天然黑玛瑙、天然黄钻、
　　　　 18K 白金、18K 黑金

Artist: Lin Hongyu（China）
Type: brooch
Material: natural coral, natural ruby, natural diamond,
natural jade, natural black agate,
natural yellow diamond,
18K platinum, 18K black gold

打藤牌、亮相、后空翻
PLAY RATTAN SHIELD, DEBUT, BACKFLIP

作者名： 林思妤（中国台湾）
类　型： 戒指
材　质： 白铜、黄铜、24K 金、树脂、漆

Artist: 　Lin Siyu（Taiwan, China）
Type: 　ring
Material: 　cupronickel, brass, 24K gold, resin, lacquer

AR 首饰
AR JEWELRY

作者名： **林羽裳（中国）**
类　型： **胸针**
材　质： **925 银、树脂**

Artist:　　Lin Yushang（China）
Type:　　　brooch
Material:　925 silver, resin

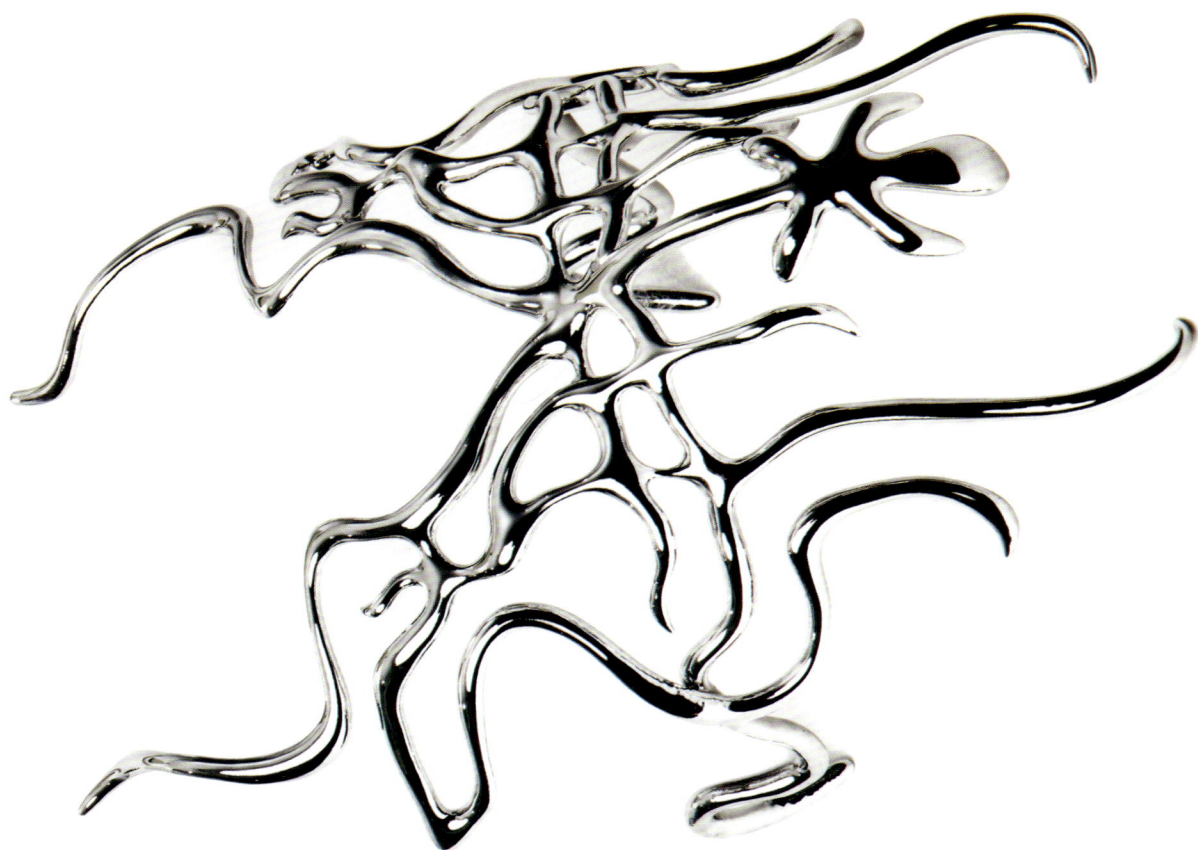

物象 —— 云
IMAGE — CLOUD

作者名： 刘建钊（中国）
类　型： 手饰
材　质： 玉、银

Artist: Liu Jianzhao（China）
Type: ornament in hand
Material: jade, silver

作者名： 刘建钊（中国）

望山
LOOK INTO THE MOUNTAIN

作者名： 刘静（中国）
类　型： 胸针、项饰
材　质： 925 银、紫光檀、珐琅

Artist: 　Liu Jing（China）
Type: 　brooch, necklace
Material: 925 silver, purple sandalwood, enamel

如金似骨
EVOLVING BONE

作者名： 刘唯芳（中国）
类　型： 项饰、耳饰
材　质： 925 银、银镀金、其他

Artist:　　Liu Weifang（China）
Type:　　 necklace, earrings
Material:　925 silver, gold-plated silver,
　　　　　 others

高屋建 "领"
BE STRATEGICALLY "SITUATED"

作者名： 刘潇女（中国）
类　型： 耳饰
材　质： 925 银、木材

Artist:　Liu Xiaonü（China）
Type:　earrings
Material:　925 silver, wood

高屋建 "领"
BE STRATEGICALLY "SITUATED"

点点
DOT DOT

作者名： 刘小奇（中国）
类　型： 胸针
材　质： 黄铜、925 银、大漆

Artist: Liu Xiaoqi（China）
Type: brooch
Material: brass, 925 silver, lacquer

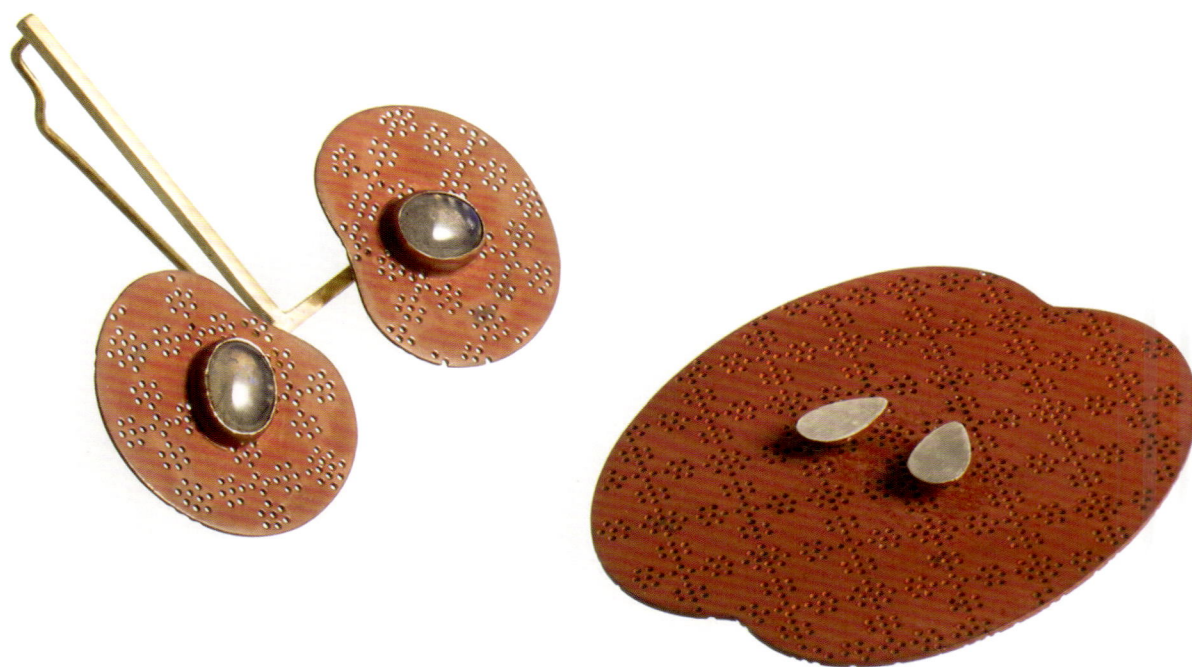

故垒
OLD RAMPARTS

作者名： 刘雪茜（中国）
类　型： 戒指
材　质： 废旧青砖、紫铜

Artist: Liu Xueqian（China）
Type: ring
Material: scrap blue bricks, copper

追金琢文
CHASE GOLD AND FINE WRITING

作者名： 刘泽慧（中国）
类　型： 胸针
材　质： 纸、金箔、树脂、
　　　　 大漆、18K 金线

Artist: Liu Zehui（China）
Type: brooch
Material: paper, gold foil, resin,
lacquer, 18K gold wire

大米基金
RICE FUND

作者名： **娄金（中国）**
类　型： **手饰**
材　质： **玉**

Artist: Lou Jin（China）
Type: ornament in hand
Material: jade

春天最后一片湖
LAST LAKE OF SPRING

作者名： **鲁霁萱（中国）**
类　型： **胸针**
材　质： **银、珐琅、钢**

Artist: Lu Jixuan（China）
Type: brooch
Material: silver, enamel, steel

容貌焦虑
APPEARANCE ANXIETY

作者名： 栾建霞（中国）
类　型： 胸针
材　质： 纯银、珐琅、不锈钢、
　　　　铜镀金、高粱

Artist:　　Luan Jianxia（China）
Type:　　　brooch
Material:　pure silver, enamel, stainless steel,
　　　　　gold-plated copper, sorghum

作者名： 栾建霞（中国）
类　型： 胸针
材　质： 纯银、珐琅、不锈钢、
　　　　铜镀金、高粱

春华秋实 —— 樱桃番茄
PRECIOUS FOOD — CHERRY TOMATOES

作者名： **罗鸿蒙（中国）**
类　型： **项链**
材　质： **黄铜**

Artist:　Luo Hongmeng（China）
Type:　necklace
Material:　brass

种子卫星
SEED SATELLITE

作者名： 吕昊洋珖（中国）
类　型： 戒指、项坠
材　质： 纯银、亚克力、不锈钢

Artist: Lü Haoyangguang（China）
Type: ring, pendant
Material: pure silver, acrylic, stainless steel

泥土？还是珍珠？
DIRT？ OR PEARL?

作者名： 马瑞希（中国）
类　型： 项饰
材　质： 亚克力、珍珠、黄铜、泥土、其他

Artist: Ma Ruixi（China）
Type: necklace
Material: acrylic, pearl, brass, soil, others

永恒的流逝
THE PASSAGE OF ETERNITY

作者名： 孟祥东（中国）
类　型： 胸针
材　质： 碧玺、锆石、18K 金、
　　　　 24K 金、925 银

Artist:　 Meng Xiangdong（China）
Type:　　brooch
Material: tourmaline, zircon, 18K gold,
　　　　　24K gold, 925 silver

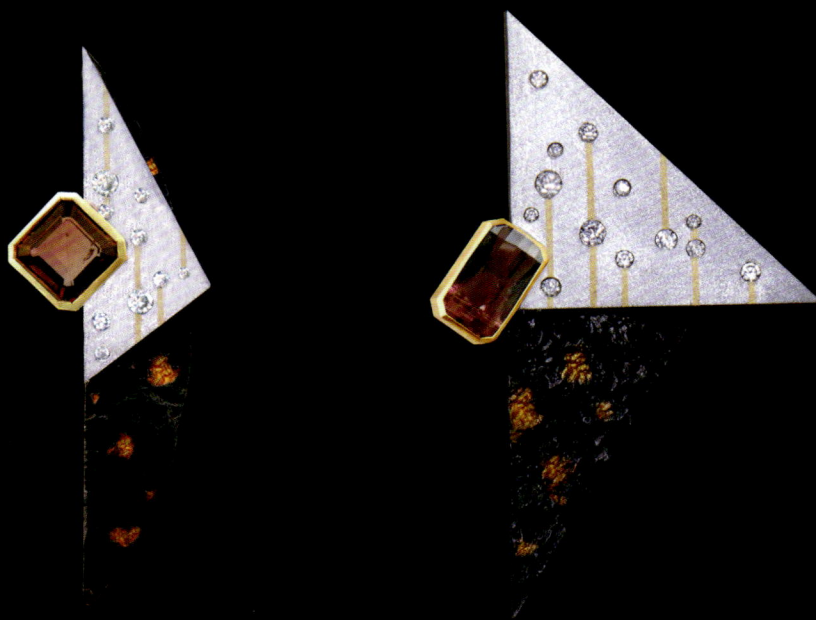

大鹏鸟
ROC

作者名： 莫文权（中国）
类　型： 吊坠
材　质： 925 银

Artist: Mo Wenquan（China）
Type: pendant
Material: 925 silver

作者名： 莫文权（中国）
类　型： 吊坠

繁花
FLOWERS

作者名： 倪献鸥（中国）
类　型： 项饰、胸针
材　质： 银、不锈钢

Artist:　　Ni Xian'ou（China）
Type:　　 necklace, brooch
Material:　silver, stainless steel

集喜
COLLECT HAPPY EVENTS

作者名： 欧阳海兵（中国）
类　型： 项饰、耳饰
材　质： 树脂、编织线、其他

Artist: Ouyang Haibing（China）
Type: necklace, earrings
Material: resin, braided wire, others

重庆脉络
CHONGQING VENATION

作者名： **潘相宜（中国）**
类　型： **胸针**
材　质： **紫铜、珐琅**

Artist:　　Pan Xiangyi（China）
Type:　　 brooch
Material: copper, enamel

记忆碎片
FRAGMENTS OF MEMORIES

作者名： **潘晓慧（中国）**
类　型： **项坠**
材　质： **陶瓷**

Artist:　　Pan Xiaohui（China）
Type:　　pendant
Material:　porcelain

作者名： **潘晓慧（中国）**

光辉岁月
GOLDEN YEARS

作者名： 潘奕吉（中国）
类　型： 项坠
材　质： 珐琅、紫铜、玻璃

Artist: Pan Yiji（China）
Type: pendant
Material: enamel, copper, glass

生计
THE LIVELIHOOD

作者名： **邱启敬（中国）**
类　型： **摆件**
材　质： **和田白玉**

Artist: Qiu Qijing（China）
Type: object
Material: hotan jade

风 · 幡 · 心之心动
WIND, FLAG, HEART'S HEARTBEAT

作者名： 屈梦楠（中国）
类　型： 胸针
材　质： 18K 绿金、999 银、925 银、真丝、
　　　　电镀聚氯乙烯、不锈钢、单向镜

Artist: 　Qu Mengnan（China）
Type: 　　brooch
Material: 　18K green gold, 999 silver, 925 silver, silk,
　　　　　electroplated polyvinyl chloride, stainless steel,
　　　　　unidirectional mirror

墨
INK

作者名： 阙城龙（中国）
类　型： 项饰、手镯、戒指
材　质： 宣纸、黑纱、麻绳、黑色线

Artist: Que Chenglong（China）
Type: necklace, bracelet, ring
Material: xuan paper, black yarn,
hemp rope, black thread

作者名： 阙城龙（中国）

大河上下
UP AND DOWN THE RIVER

作者名： 任海明（中国）
类　型： 戒指
材　质： 18K 金、黄铜、锆石、尼龙

Artist: 　Ren Haiming（China）
Type: 　　ring
Material: 　18K gold, brass, zircon, nylon

生生不息
CIRCLE OF LIFE

作者名： 任雪倪（中国）
类　型： 项饰
材　质： 金箔、紫铜、螺钿

Artist: Ren Xueni（China）
Type: necklace
Material: gold foil, copper,
mother-of-pearl

绪合计划 —— 分为男款和女款
XUHE PLAN — DIVIDED INTO MEN'S AND WOMEN'S STYLE

作者名：沙睿琬、郑月朗、陈缘圆（中国）
类　型：手镯
材　质：虚拟影像、其他

Artist:　Sha Ruiwan, Zheng Yuelang,
　　　　Chen Yuanyuan（China）
Type:　bracelet
Material:　virtual image, others

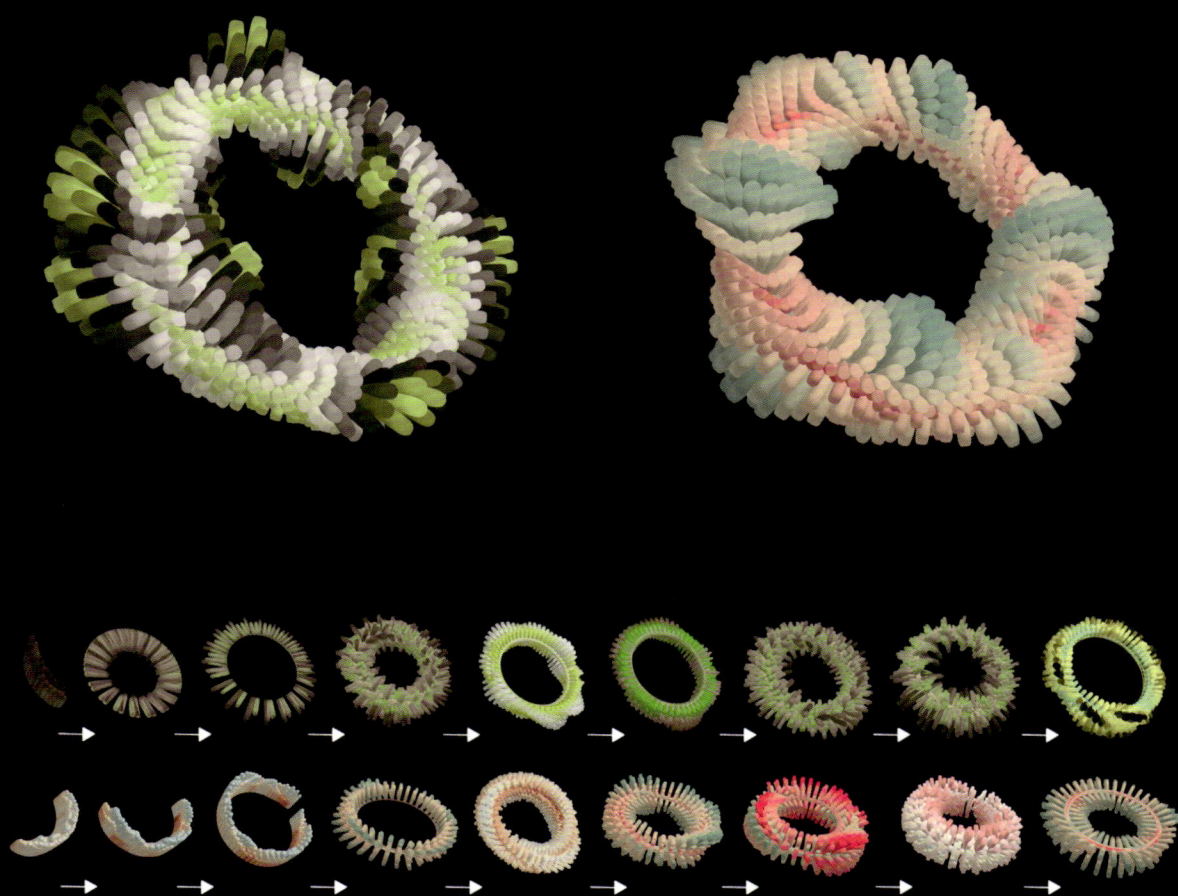

花开 —— 芳华
FLOWERS BLOOM — FANGHUA

作者名：　邵萍（中国）
类　型：　项饰、胸针
材　质：　18K 金、钛、珍珠、钻石、沙弗莱石

Artist:　　Shao Ping（China）
Type:　　necklace, brooch
Material:　18K gold, titanium, pearl, diamond, tsavorite

消失
DISAPPEARING

作者名： 石芮宁（中国）
类　型： 项饰
材　质： 925 银、铁、银箔、尼龙绳

Artist:　　Shi Ruining（China）
Type:　　 necklace
Material:　925 silver, iron,
　　　　　 silver foil, nylon rope

象外之象
UNEXPECTED SIGHT

作者名： 宋晓薇、汪润东（中国）
类　型： 项坠、手链
材　质： 925 银、珐琅、锆石、18K 金

Artist:　　Song Xiaowei, Wang Rundong（China）
Type:　　 pendant, bracelet
Material:　925 silver, enamel, zircon, 18K gold

收集感知的三个盒子
THREE BOXES OF
COLLECTING PERCEPTION

作者名： 孙汇泽（中国）
类　型： 项饰
材　质： 黄铜、925 银、钢、尼龙、其他

Artist: Sun Huize（China）
Type: necklace
Material: brass, 925 silver, steel, nylon, others

反：前言
ANTI: FOREWORD

作者名： 孙铭瑞（中国）
类　型： 胸针
材　质： 塑料、925 银、钢

Artist: Sun Mingrui（China）
Type: brooch
Material: plastic, 925 silver, steel

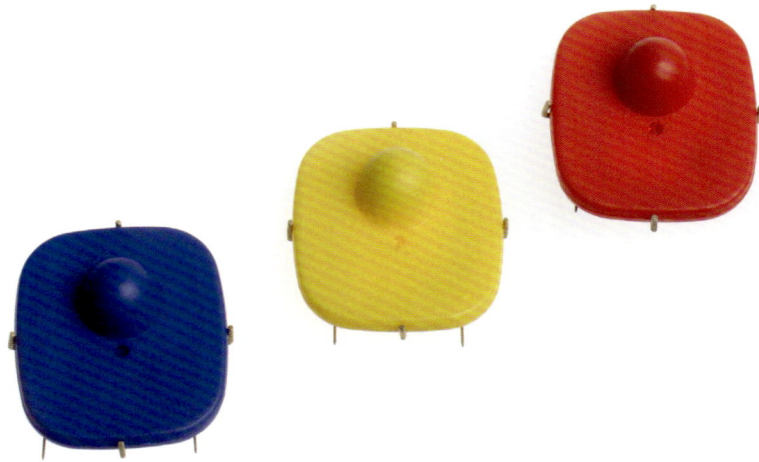

足迹
FOOTPRINTS

作者名： 孙铭燕（中国）
类　型： 胸针
材　质： 925 银、皮革边角料、其他

Artist: Sun Mingyan（China）
Type: brooch
Material: 925 silver, leather scraps, others

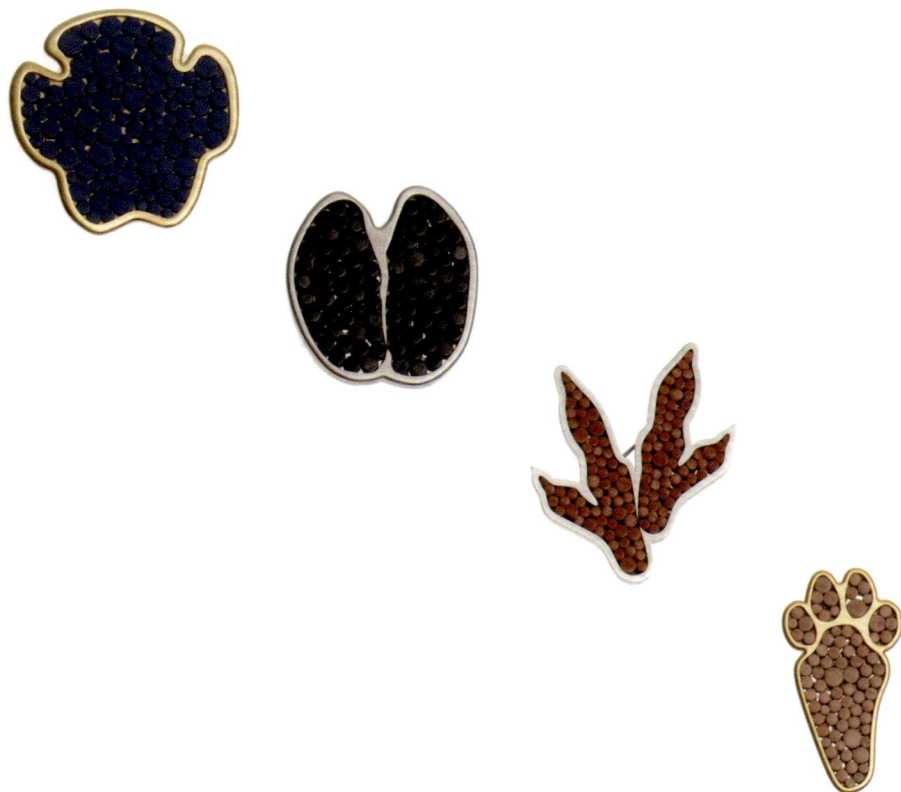

岁月如歌
TIME IS LIKE A SONG

作者名： 孙秋爽（中国）
类　型： 项饰
材　质： 纯银、24K 金、树脂

Artist: Sun Qiushuang（China）
Type: necklace
Material: pure silver, 24K gold, resin

神秘饰博志
ABOUT MYSTERIOUS JEWELRY

作者名： 谭楚童（中国）
类　型： 项饰、胸针、摆件
材　质： 银、金、大漆、黄铜、
　　　　 紫铜、化石、宝石、珍珠

Artist: Tan Chutong（China）
Type: necklace, brooch, object
Material: silver, gold, lacquer, brass,
copper, fossil, gemstone, pearl

作者名： 谭楚童（中国）
类　型： 项饰、胸针、摆件

都市生存焦虑
URBAN SURVIVAL ANXIETY

作者名： 唐含芝（中国）
类　型： 胸针
材　质： 黄铜、陶瓷、玻璃、砖块、
　　　　 铁锈、墙皮、水泥、建筑废料

Artist:　Tang Hanzhi（China）
Type:　brooch
Material:　brass, ceramic, glass, brick, rust,
　　　　　wall covering, cement, construction waste

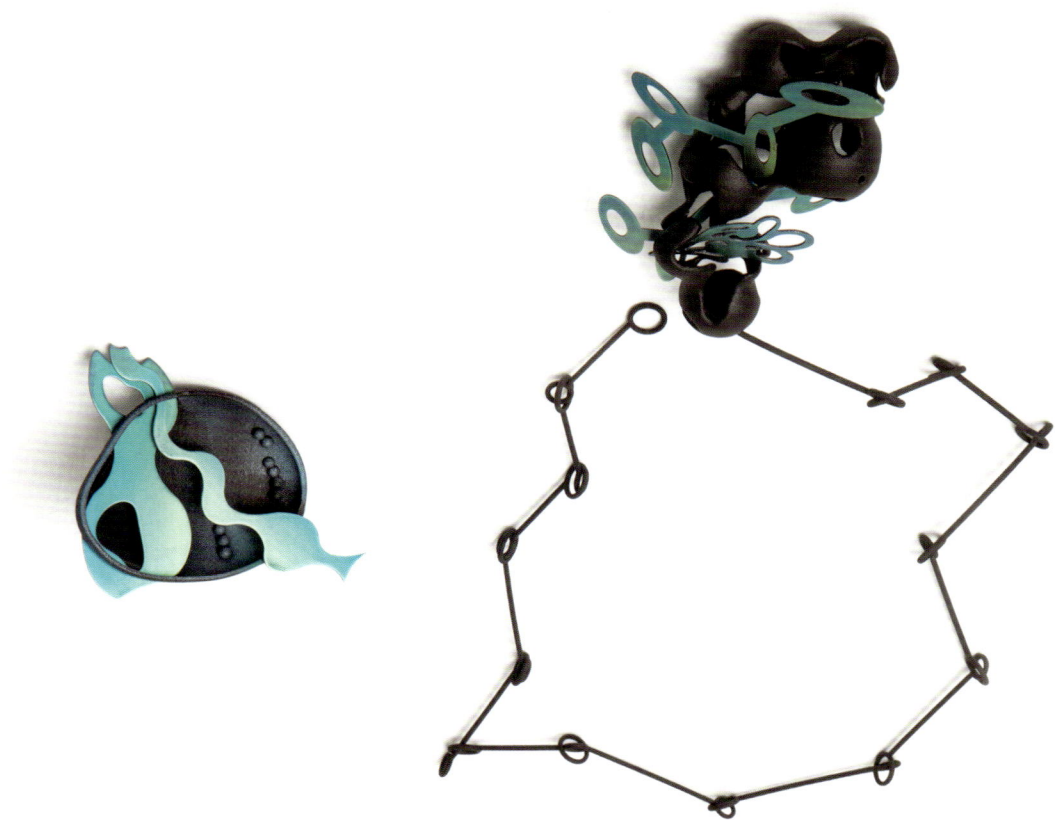

母体机器
MATRIX OF MACHINE

作者名： 王晗（中国）
类 型： 项饰、胸针
材 质： 钛、尼龙、电子元件、其他

Artist: Wang Han（China）
Type: necklace, brooch
Material: titanium, nylon,
electronic components, others

蝶舞
BUTTERFLY

作者名： 田伟玲（中国）
类 型： 项饰
材 质： 纯银

Artist: Tian Weiling（China）
Type: necklace
Material: pure silver

Material: emerald, diamond, aquamarine, 18K gold
Type: brooch
Artist: Wang Haorui, Sun Ying (China)

材质：祖母绿、钻石、海蓝宝、18K 金
类型：胸针
作者名：王浩睿、孙颖（中国）

LOTUS LEAF

荷叶

Material: 925 silver, zircon,
pearl, enamel

Type: pendant, brooch

Artist: Wang Haoyu (China)

材质: 925银、锆石、珍珠、珐琅

类型: 吊坠、胸针

作者名: 王皓宇（中国）

MEMOIR

回忆录

寂之声·冬韵
THE SOUND OF WABI-SABI · WINTER RHYME

作者名： 王敬（中国）
类 型： 胸针
材 质： 925 银、珐琅、金

Artist: Wang Jing（China）
Type: brooch
Material: 925 silver, enamel, gold

营养素 –1
NUTRIENTS–1

作者名： 王平（中国）
类　型： 项饰、戒指
材　质： 维生素、亚克力圆球盒、
　　　　 鱼线、金属项圈、其他

Artist: 　Wang Ping（China）
Type: 　　necklace, ring
Material: 　vitamins, acrylic ball box,
　　　　　 fishing line, metal collar, others

作者名： 王平（中国）

浮动花园
FLOATING GARDEN

作者名： 王茜（中国）
类　型： 胸针、项链
材　质： 氟铝石膏、925 银

Artist:　　Wang Xi（China）
Type:　　brooch, necklace
Material:　fluoroaluminum gypsum,
　　　　　925 silver

冲和为德
INTEGRATING THE CONDUCT OF DIFFERENT THINGS

作者名： 王绍伟（中国）
类　型： 项饰
材　质： 紫铜、白铜、珍珠、
　　　　木材、不锈钢

Artist:　　Wang Shaowei（China）
Type:　　 pendant
Material: copper, cupronickel, pearl,
　　　　　wood, stainless steel

无题
UNTITLED

作者名： 王圣临（中国）
类　型： 耳饰
材　质： 18K 金、摩根石、风信子、
　　　　 钻石、海蓝宝、养殖鹅羽

Artist:　　 Wang Shenglin（China）
Type:　　 earrings
Material:　 18K gold, morganite, hyacinth, diamond,
　　　　　 aquamarine, cultured goose feather

印象山茶花
THE IMPRESSION OF CAMELLIA

作者名： 王书利（中国）
类　型： 项坠
材　质： 纯银、珍珠

Artist: Wang Shuli（China）
Type: necklace
Material: pure silver, pearl

心心相印
PADA

作者名： 王亚芳（中国）
类　型： 胸针
材　质： 大漆、黄铜、钢、蛋壳、其他

Artist: Wang Yafang（China）
Type: brooch
Material: lacquer, brass, steel, eggshell, others

荡漾
RIPPLE

作者名： 王莹（中国）
类 型： 胸针
材 质： 925 银、螺钿、碧玺、橄榄石

Artist: Wang Ying（China）
Type: brooch
Material: 925 silver, mother-of-pearl,
tourmaline, peridot

万物生
ALL THINGS ARE BORN

作者名： 王玉囡（中国）
类　型： 项饰、戒指
材　质： 925 银、珐琅、红铜、珍珠、银箔

Artist: Wang Yunan（China）
Type: necklace, ring
Material: 925 silver, enamel,
red copper, pearl, silver foil

无题
UNTITLED

作者名: **王泽丹（中国）**
类　型: **戒指**
材　质: **足金、钻石**

Artist: Wang Zedan（China）
Type: ring
Material: pure gold, diamond

穿过花园的大门
THROUGH THE GARDEN GATE

作者名： 汪正虹（中国）
类　型： 胸针
材　质： 925 银、棉线、木材

Artist: Wang Zhenghong（China）
Type: brooch
Material: 925 silver, cotton thread, wood

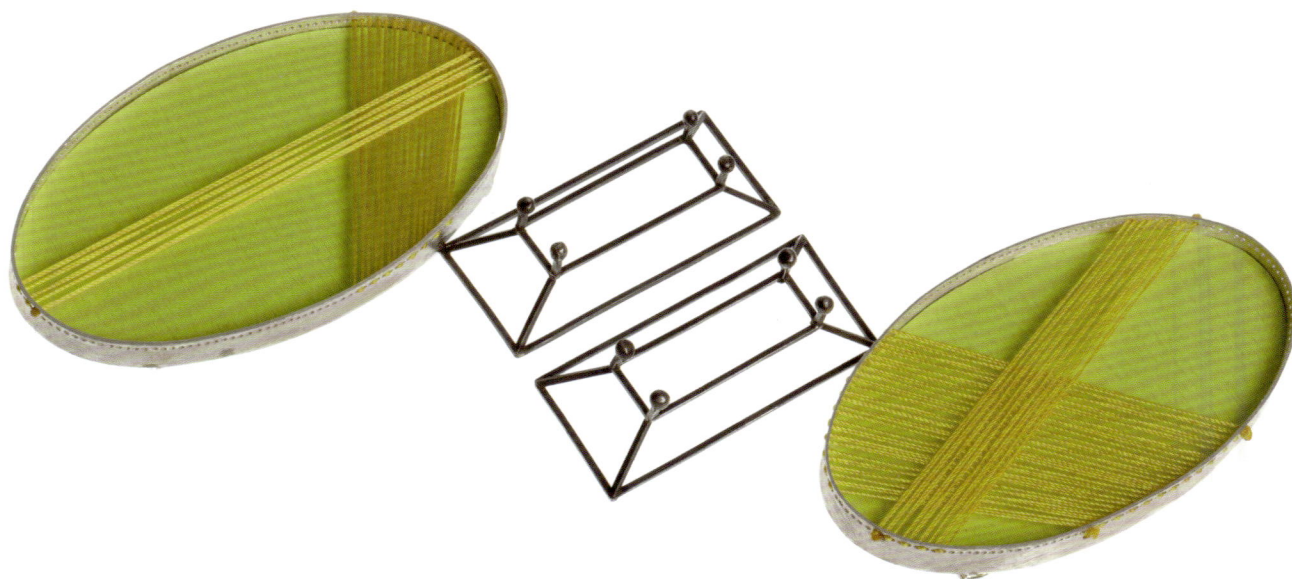

红线与壶
POT WITH RED THREAD

作者名： **魏子欣（中国）**
类　型： **摆件**
材　质： **纯银、钢、棉线**

Artist: *Wei Zixin（China）*
Type: *object*
Material: *pure silver, steel, cotton thread*

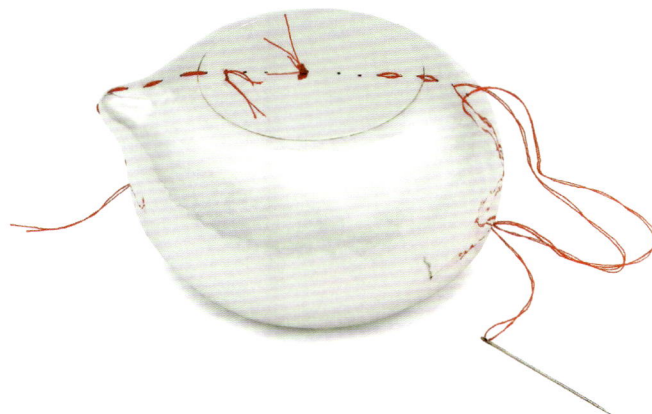

疫时光
THE DAY OF THE EPIDEMIC

作者名： **吴二强（中国）**
类　型： **项饰**
材　质： **鸡骨头、电话线、其他**

Artist: *Wu Erqiang（China）*
Type: *necklace*
Material: *chicken bones, telephone cord, others*

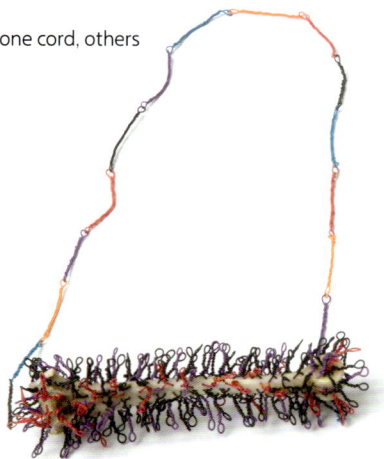

微风·存档
BREEZE·FILE

作者名： 吴芳（中国）
类 型： 项坠
材 质： 纯银、塑料、缝纫线、其他

Artist: Wu Fang（China）
Type: pendant
Material: silver, plastic, sewing thread, others

乌江之息
BREATH OF WUJIANG RIVER

作者名： 吴靓怡（中国）
类　型： 项饰
材　质： 925 银、树脂、锆石、珐琅

Artist: Wu Jingyi（China）
Type: necklace
Material: 925 silver, resin, zircon, enamel

鼻炎患者之痛
RHINITIS PATIENTS' PAIN

作者名： 吴人杰（中国）
类　型： 胸针
材　质： 纯银、珍珠

Artist: Wu Renjie（China）
Type: brooch
Material: pure silver, pearl

J.OIE —— 想象探索之旅
J.OIE — IMAGINARY EXPLORATION JOURNEY

作者名： 吴颐藻（中国）
类　型： 胸针、耳饰
材　质： 数字协同共创作品、其他

Artist: Wu Yizao（China）
Type: brooch, earrings
Material: digital collaborative co-creation works, others

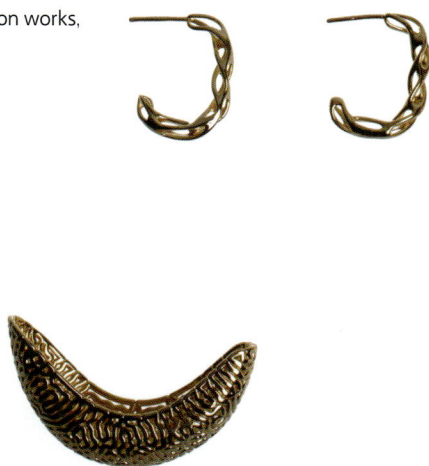

冬 · 蕴系列之簌
WINTER STORAGE SERIES – SU

作者名： 吴越卓（中国）
类　型： 胸针
材　质： 钛、925 银、24K 金、大漆、牛角

Artist: Wu Yuezhuo（China）
Type: brooch
Material: titanium, 925 silver, 24K gold, lacquer, horns

未来佛
FUTURE BUDDHA

作者名： 吴正军（中国）
类　型： 项坠
材　质： 925 银、黄铜

Artist: Wu Zhengjun（China）
Type: pendant
Material: 925 silver, brass

未来之狐
FUTURE FOX

作者名： **谢馥蔚（中国）**
类　型： 项坠、胸针
材　质： 925 银、锆石

Artist:　　Xie Fuwei（China）
Type:　　　pendant, brooch
Material:　925 silver, zircon

作者名： 解雅婷（中国）
类　型： 胸针
材　质： 玉线、翡翠、其他

Artist:　　Xie Yating（China）
Type:　　brooch
Material:　jade thread, jadeite, others

敦煌搜神
SEARCH IMMORTALS IN DUNHUANG

作者名： 熊芏芏（中国）
类　型： 胸针
材　质： 白玉、黑青玉、钻石、18K 金、
　　　　 18K 白金、绿松石、青金石、钛

Artist:　　Xiong Dudu（China）
Type:　　 brooch
Material:　white jade, black sapphire, diamond,
　　　　　18K gold, 18K platinum, turquoise,
　　　　　lapis lazuli, titanium

姿态与能量 1
GESTURE AND ENERGY 1

作者名： 许晨茜（中国）
类　型： 胸针
材　质： 钢、不锈钢、珐琅

Artist:　　Xu Chenxi（China）
Type:　　 brooch
Material:　steel, stainless steel, enamel

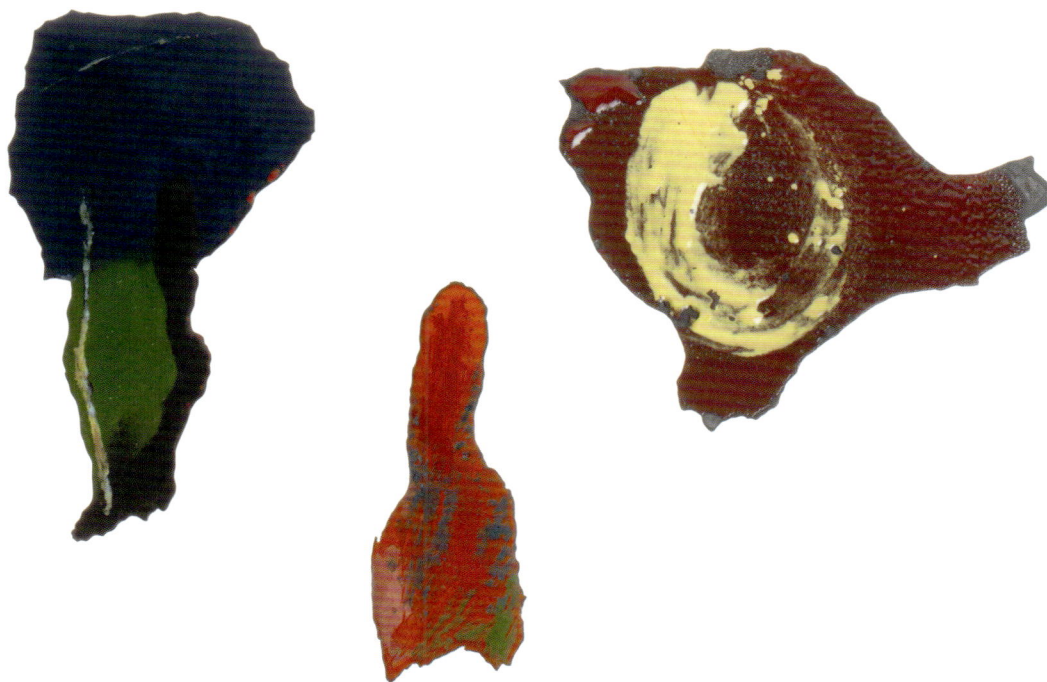

复圆，复原
RESTORE ROUND SHAPE—REVERSION

作者名： 薛菲（中国）
类　型： 项饰、手镯
材　质： 黄铜、翡翠

Artist: Xue Fei（China）
Type: necklace, bracelet
Material: brass, emerald

良夜·金属重工多用置物挂饰项链 1
METAL NIGHT–
METAL HEAVY INDUSTRY MULTI–
PURPOSE PENDANT NECKLACE 1

作者名： 闫丹婷（中国）
类 型： 吊坠
材 质： 黄铜、925 银、不锈钢

Artist: Yan Danting（China）
Type: pendant
Material: brass, 925 silver, stainless steel

未来"文、物"
FUTURE "CULTURE AND OBJECTS"

作者名： 闫政旭（中国）
类 型： 胸针
材 质： 纯银、925 银、珐琅

Artist: Yan Zhengxu（China）
Type: brooch
Material: pure silver, 925 silver, enamel

云
CLOUD

作者名： 杨漫（中国）
类　型： 胸针
材　质： 925 银、银镀亮银、
　　　　 银镀 18K 金、银镀电玫瑰金

Artist:　　Yang Man（China）
Type:　　 brooch
Material:　925 silver, bright silver-plated silver,
　　　　　 18K gold-plated silver,
　　　　　 rose gold-plated silver

作者名： 杨漫（中国）

静思妙悟
MEDITATION AND ENLIGHTENMENT

作者名： 杨乾（中国）
类　型： 吊坠
材　质： 紫绿玛瑙

Artist:　Yang Qian（China）
Type:　pendant
Material:　purple green agate

炫（一）
DAZZLING (ONE)

作者名： **杨小庆（中国）**
类　型： **项饰、耳饰、手链**
材　质： **玻璃、黄铜、其他**

Artist: Yang Xiaoqing（China）
Type: necklace, earrings, bracelet
Material: glass, brass, others

空 · 缺 2-1
EMPTY & LACK 2-1

作者名： 杨中雄（中国）
类　型： 戒指
材　质： 纯银、陶瓷

Artist: Yang Zhongxiong（China）
Type: ring
Material: silver, porcelain

虚实混境
MIXED REALM OF VIRTUAL AND REAL

作者名： 姚遥（中国）
类　型： 胸针
材　质： 紫铜、锆石

Artist: Yao Yao（China）
Type: brooch
Material: copper, zircon

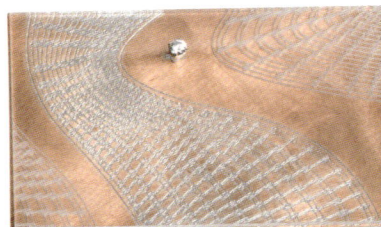

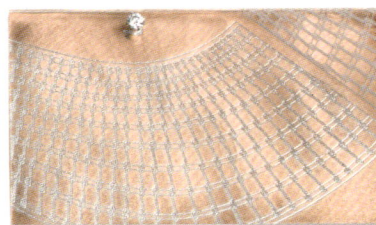

万物生
EVERYTHING IN EXISTENCE

作者名: 曾嫚（中国）
类 型: 项饰
材 质: 925 银、紫晶原石

Artist: Zeng Man（China）
Type: necklace
Material: 925 silver, amethyst raw stone

配角，主角
BIT PART, MAIN PART

作者名: 曾若琅（中国）
类 型: 项饰、胸针
材 质: 黄铜

Artist: Zeng Ruolang（China）
Type: necklace, brooch
Material: brass

孵化器 —— 深海
HATCHERY — ABYSSAL ZONE

作者名： 詹瞻（中国）
类　型： 戒指
材　质： 925 银、珐琅

Artist: Zhan Zhan（China）
Type: ring
Material: 925 silver, enamel

永绽
BLOOM FOREVER

作者名： 张楚楚（中国）
类　型： 胸针
材　质： 人造宝石、锆石、树脂、玻璃、
　　　　 亚克力、白铜、棉线、鱼线、
　　　　 金线、棉布、颜料、其他

Artist: Zhang Chuchu（China）
Type: brooch
Material: artificial gemstone, zircon, resin, glass,
acrylic, white copper, cotton thread,
fishing thread, gold thread,
cotton cloth, paint, others

生如夏花系列
FLOWER OF INSECTS

作者名： **张慧敏（中国）**
类　型： **戒指**
材　质： **22K 金**

Artist:　　Zhang Huimin（China）
Type:　　ring
Material:　22K gold

二十四天心
TWENTY FOUR DAYS HEART

作者名： 张凡（中国）
类　型： 项饰
材　质： 珍珠、紫铜鎏银、
　　　　紫铜鎏金、AR 交互设备

Artist: Zhang Fan（China）
Type: necklace
Material: pearl, copper gilt silver,
copper gilt gold,
AR interactive device

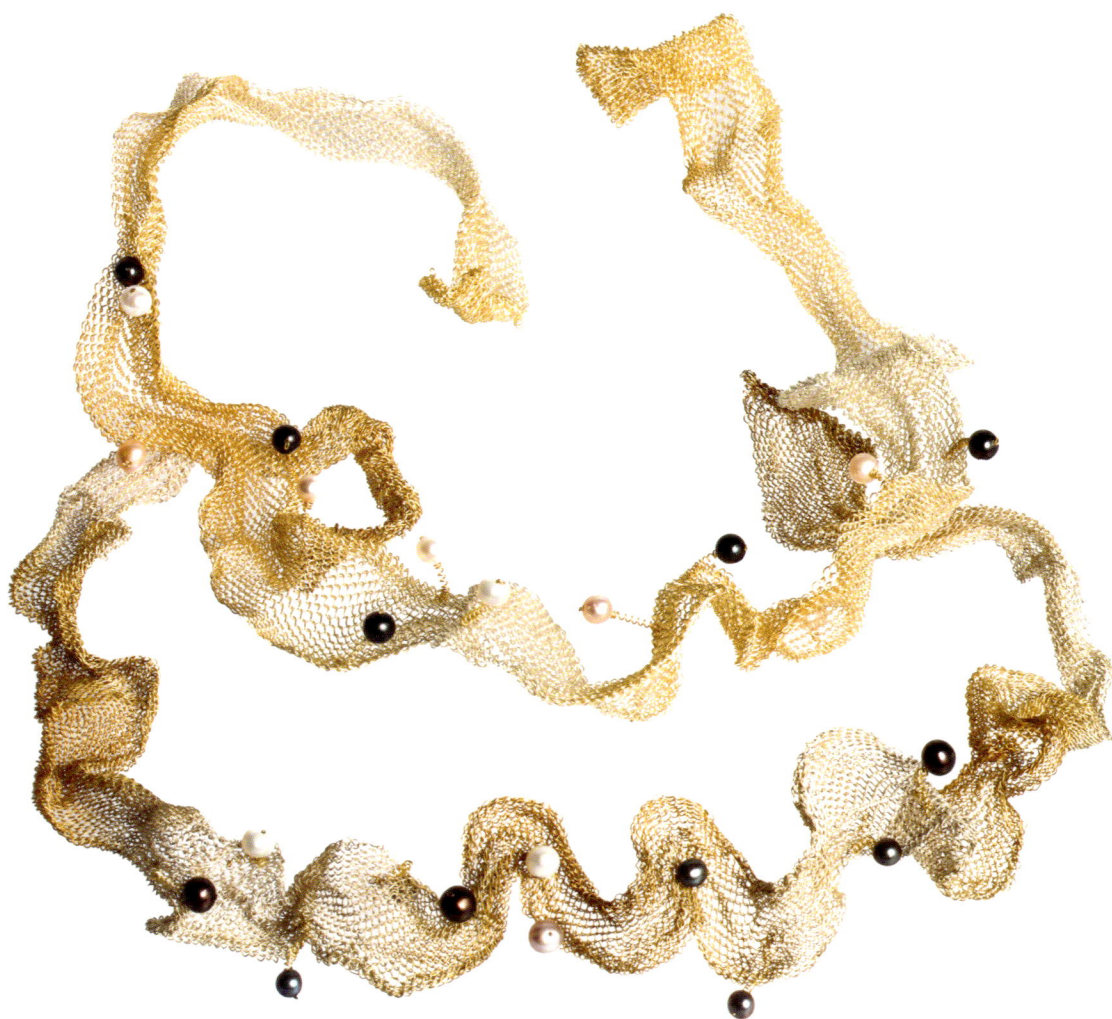

碳基遗迹
SITES OF CARBON —BASED

作者名： 张楠楠（中国）
类　型： 项饰、胸针、耳饰
材　质： 925 银、珍珠、祖母绿、布面油画

Artist:　　Zhang Nannan（China）
Type:　　 necklace, brooch, earrings
Material:　925 silver, pearl, emerald, oil on canvas

蚕语
COCOONING

作者名： 张思秋（中国）
类　型： 胸针
材　质： 纯银

Artist:　　Zhang Siqiu（China）
Type:　　 brooch
Material:　pure silver

万物生系列
EVERYTHING GROWS SERIES

作者名： 张雯迪（中国）
类　型： 胸针
材　质： 纯银、珐琅、火山石、其他

Artist:　　Zhang Wendi（China）
Type:　　 brooch
Material:　pure silver, enamel,
　　　　　volcanic stone, others

涅槃之花：重生的宝石
NIRVANA FLOWER:
A REBORN GEMSTONE

作者名： 张压西、申子叶（中国）
类　型： 项坠、戒指
材　质： 玻璃

Artist: Zhang Yaxi, Shen Ziye（China）
Type: necklace, ring
Material: glass

落日珊瑚
SUNSET CORAL

作者名： 张莹（中国）
类　型： 手镯
材　质： 18K 金、彩色蓝宝石、钻石、红宝石

Artist: Zhang Ying（China）
Type: bracelet
Material: 18K gold, fancy sapphires,
diamond, ruby

啊咧呀啰 —— 童谣
AH LIE YA LUO — NURSERY RHYME

作者名： 张玉（中国）
类　型： 胸饰
材　质： 树脂、黄铜镀 14K 金

Artist:　　Zhang Yu（China）
Type:　　　brooch
Material:　resin, 14K gold plated on brass

契约
CONTRACT

作者名： **章美薇（中国）**
类　型： **戒指**
材　质： **925 银、珐琅**

Artist:　　Zhang Meiwei（China）
Type:　　ring
Material:　925 silver, enamel

黔南手迹 · 叠 1
HANDWRITING OF SOUTHERN GUIZHOU · STACK 1

作者名： 章雪莶（中国）
类　型： 手镯、胸针
材　质： 纯银、南红玛瑙

Artist:　　Zhang Xueying（China）
Type:　　 bracelet, brooch
Material:　pure silver, southern red agate

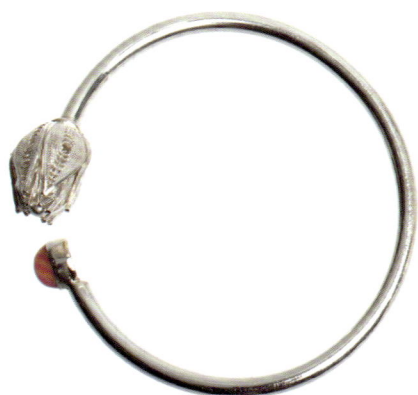

作者名： 章雪莶（中国）

拥抱一只刺猬
HUG A HEDGEHOG

作者名： 赵洁（中国）
类　型： 项饰、手镯、戒指
材　质： 大漆、白铜、蓝宝石、
　　　　 海蓝宝、羊毛毡、其他

Artist: Zhao Jie（China）
Type: necklace, bracelet, ring
Material: lacquer, white copper, sapphire,
aquamarine, wool felt, others

幸福的爱
PEACEFUL LOVE

作者名： 赵世笺（中国）
类　型： 胸针
材　质： 925 银、珐琅

Artist: Zhao Shijian（China）
Type: brooch
Material: 925 silver, enamel

十年
TEN YEARS

作者名： 赵祎（中国）
类　型： 项饰
材　质： 银、木材、大漆

Artist:　　Zhao Yi（China）
Type:　　 necklace
Material:　 silver, wood, lacquer

浮山掠影
GLIMPSES OF FLOATING MOUNTAIN

作者名： 郑静、赵欣然（中国）
类　型： 胸针、戒指
材　质： 925 银、3D 打印树脂

Artist:　　Zheng Jing, Zhao Xinran（China）
Type:　　　brooch, ring
Material:　925 silver, 3D printing resin

作者名： 郑静、赵欣然（中国）
类　型： 胸针、戒指

山谷里的声音
VOICES IN THE VALLEY

作者名： 郑妍芳（中国）
类　型： 耳饰
材　质： 925银、纯银、珍珠、
　　　　 竹、鱼鳞、其他

Artist: Zheng Yanfang（China）
Type: earrings
Material: 925 silver, pure silver, pearl,
bamboo, fish scales, others

元
YUAN

作者名： 周芳（中国）
类　型： 戒指
材　质： 大漆、金箔、其他

Artist: Zhou Fang（China）
Type: ring
Material: lacquer, gold leaf, others

福从何处生
WISH YOU ALL WELL

作者名： 周若雪（中国）
类　型： 胸针
材　质： 钢、银镀金、纸、金箔

Artist:　　Zhou Ruoxue（China）
Type:　　　brooch
Material:　steel, gold plated on silver,
　　　　　　paper, gold foil

漆·书
LACQUER·BOOK

作者名： 周子琪（中国）
类　型： 胸针
材　质： 大漆、树脂

Artist:　　Zhou Ziqi（China）
Type:　　　brooch
Material:　lacquer, resin

玉望
DESIRE OF JADE

作者名： 朱鹏飞（中国）
类　型： 项坠
材　质： 925 银、白玉、电机、电池

Artist: Zhu Pengfei（China）
Type: pendant
Material: 925 silver, white jade, motor, battery

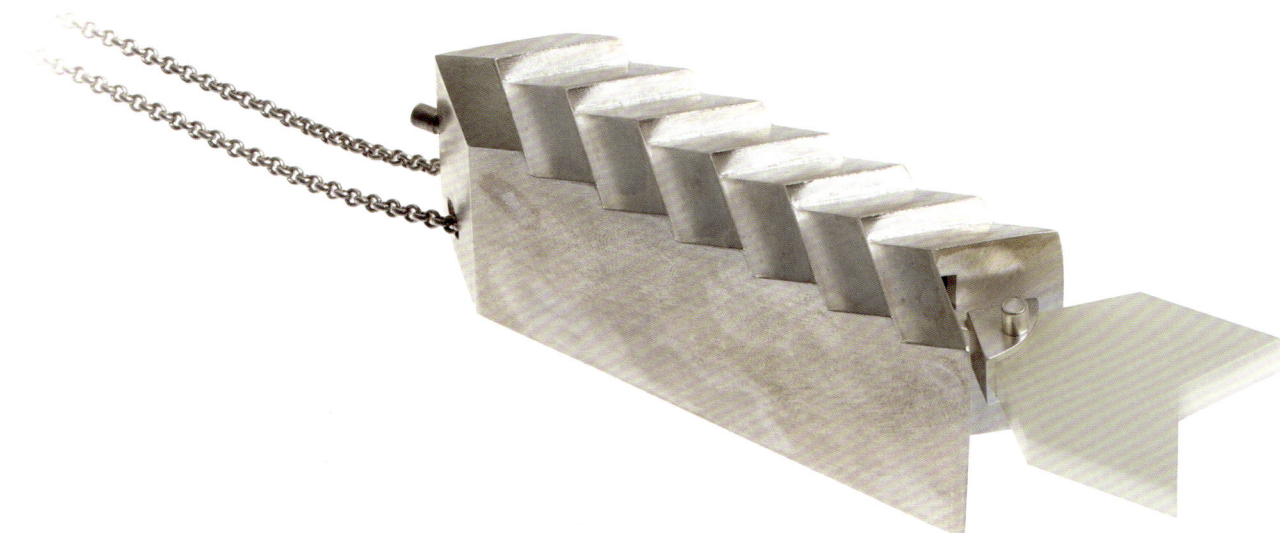

彩盒子 II
COLOURFUL BOX II

作者名： 庄冬冬（中国）
类 型： 胸针
材 质： 大漆、螺钿、塑料、尼龙

Artist: Zhuang Dongdong（China）
Type: brooch
Material: lacquer, mother-of-pearl, plastic, nylon

作者名： 邹宁馨、张婧卓、赵思雨、
马玥、唐佳琪（中国）

类　型： 胸针

材　质： 银

Artist: Zou Ningxin, Zhang Jingzhuo, Zhao Siyu,
Ma Yue, Tang Jiaqi（China）

Type: brooch

Material: silver

品牌专区
BRAND ZONE

绣合欢
EMBROIDERY OF SILK TREE

作者名： **百泰集团（中国）**
类　型： **项饰**
材　质： **18K 金**

Artist: **BATAR GROUP（China）**
Type: necklace
Material: 18K gold

蝶舞
DANCE OF BUTTERFLY

作者名： 辫子珠宝（中国）
类　型： 胸针
材　质： 红宝石、黑钻、白钻、
　　　　贝母、黑玛瑙、18K 金

Artist:　　Braid's Story Jewelry（China）
Type:　　 brooch
Material:　ruby, black diamond, white diamond,
　　　　　mother-of-pearl, black agate, 18K gold

菜百传世 · 轻舞飞扬赋竹
DISHES HANDED DOWN FROM GENERATION
TO GENERATION · DANCING LIGHTLY AND
FLUTTERING WITH BAMBOO

作者名： 菜百首饰（中国）
类 型： 耳饰、项链、戒指
材 质： 足金、钻石

Artist: **CAIBAI JEWELRY**（China）
Type: earrings, pendant, ring
Material: pure gold, diamond

双鱼探花
DOUBLE FISHES EXPLORING FLOWER

作者名： 呈元珠宝设计（中国）
类　型： 胸针
材　质： 翡翠、18K 金、钻石

Artist:　　Chengyuan Jewelry Design（China）
Type:　　　brooch
Material:　Emerald, 18K gold, diamond

仙鹤
CRANE

作者名： 大树珠宝（中国）
类　型： 胸针
材　质： 18K 金、钻石、和田玉籽料

Artist: EVERGREEN JEWELRY（China）
Type: brooch
Material: 18K gold, diamond.
Hetian jade seed material

作者名： 大树珠宝（中国）

向日葵
SUNFLOWER

作者名： 广州市皇誉族珠宝首饰有限公司（中国）
类　型： 胸针
材　质： 18K 金、沙弗莱石

Artist:　Guangzhou Huangyuzu Jewelry Co., Ltd（China）
Type:　brooch
Material:　18K gold, tsavorite

花边魅影
BLOSSOM OF SHADE

作者名： 广州市天丰珠宝有限公司（中国）
类　型： 项饰、胸针
材　质： 925 银、锆石

Artist:　　Guangzhou Tianfeng Jewelry Co., Ltd.（China）
Type:　　 necklace, brooch
Material:　925 silver, zircon

青绿鸿运龙柱
UNFADING GREEN HONGYUN DRAGON PILLAR

作 者 名： 国泉金业（北京）文化股份有限公司（中国）
类　型： 吊坠
材　质： 足金 999、珐琅

Artist: Guoquan Gold (Beijing) Culture Co., Ltd.（China）
Type: pendant
Material: 999 gold, enamel

祥云
AUSPICIOUS CLOUDS

作者名： 金城皇（中国）
类　型： 项坠
材　质： 18K 金

Artist: Jinchenghuang（China）
Type: necklace
Material: 18K gold

古韵六期系列
SIX-ISSUE SERIES OF ANCIENT RHYME

作者名： 老庙黄金（中国）
类　型： 项坠、手链
材　质： 足金

Artist:　　Laomiao Jewelry（China）
Type:　　　pendant, bracelet
Material:　pure gold

永恒
ETERNAL

作者名： 梦金园黄金珠宝集团（中国）
类　型： 项饰
材　质： 足金、珐琅

Artist:　　MOKINGRAN JEWELRY（China）
Type:　　 necklace
Material:　pure gold, enamel

"链尚" 东西
"CHAIN SHANG" EAST AND WEST

作者名： NK 东西宫（中国）
类　型： 项链、手镯
材　质： 和田玉、树纹雕金、紫水晶、
　　　　彩宝、乌金、珐琅

Artist:　 NEWWWLOOK Jewelry & Gift（China）
Type:　　necklace, bracelet
Material:　Hotan jade, tree pattern and gold carving,
　　　　　amethyst, colorful gemstones,
　　　　　black gold, enamel

活力长隆
DYNAMIC CHANGLONG

作者名：　全球智造（广州）珠宝科技有限公司（中国）
类　型：　胸针、吊坠
材　质：　18K金、红宝石、沙弗莱石、钻石

Artist:　Global Smart Manufacturing (Guangzhou) Jewelry Technology Co., Ltd （China）
Type:　brooch, pendant
Material:　18K gold, ruby, tsavorite, diamond

梦蝶
DREAM BUTTERFLY

作者名： 上海老凤祥珐琅艺术有限公司（中国）
类 型： 器皿、胸针
材 质： 银、珐琅

Artist: Shanghai Laofengxiang Enamel Art Co., Ltd（China）
Type: object, brooch
Material: silver, enamel

潮极神兽系列
CHAOJI MYTHICAL BEAST SERIES

作者名： 深圳峰汇珠宝首饰有限公司（中国）
类　型： 项坠、手链
材　质： 5D 电黑

Artist:　　 Shenzhen Foreway Jewelry Co., Ltd.（China）
Type:　　　 necklace, bracelet
Material:　 5D electric black

惊蛰
AWAKENING OF INSECTS

作者名：　赵环宇、王伟（正德东奇珠宝）（中国）
类　型：　胸针
材　质：　18K 金、钻石、珐琅

Artist:　Zhao Huanyu, Wang Wei（Zhengde Dongqi Jewelry）（China）
Type:　brooch
Material:　18K gold, diamond, enamel